# HOW TO READ
# WATER

ALSO BY TRISTAN GOOLEY

*The Lost Art of Reading Nature's Signs*

*The Natural Navigator*

*The Natural Explorer*

*How to Connect with Nature*

# HOW TO READ
# WATER

## Clues and Patterns
## *from* Puddles *to the* Sea

*Learn to Gauge Depth • Navigate • Forecast Weather*
*and Make Other Predictions with Water*

# TRISTAN GOOLEY

### Illustrations by Neil Gower

THE EXPERIMENT

NEW YORK

# To the Gs, Ks, Ms, and Bs

HOW TO READ WATER: *Clues and Patterns from Puddles to the Sea*
Copyright © 2016 by Tristan Gooley
Illustrations copyright © 2016 by Neil Gower
Photographs copyright © 2016 by Tristan Gooley
Additional photo credits on page 377

Originally published in the UK by Sceptre in 2016.
First published in North America by The Experiment, LLC, in 2016.

The Experiment, LLC
220 East 23rd Street, Suite 600
New York, NY 10010-4674
theexperimentpublishing.com

Many of the designations used by manufacturers and sellers to distinguish their products are claimed as trademarks. Where those designations appear in this book and The Experiment was aware of a trademark claim, the designations have been capitalized.

The Experiment's books are available at special discounts when purchased in bulk for premiums and sales promotions as well as for fundraising or educational use. For details, contact us at info@ theexperimentpublishing.com.

Library of Congress Cataloging-in-Publication Data

Names: Gooley, Tristan, author. | Gower, Neil, illustrator.
Title: How to read water : clues and patterns from puddles to the sea / Tristan Gooley ; illustrations by Neil Gower.
Description: New York, NY : The Experiment, 2016. | Includes bibliographical references and index. | Description based on print version record and CIP data provided by publisher; resource not viewed.
Identifiers: LCCN 2016021060 (print) | LCCN 2016016408 (ebook) | ISBN 9781615193592 (e-book) | ISBN 9781615193585 (cloth)
Subjects: LCSH: Water.
Classification: LCC GB671 (print) | LCC GB671 .G66 2016 (ebook) | DDC 551.46--dc23
LC record available at https://lccn.loc.gov/2016021060

ISBN 978-1-61519-358-5
Ebook ISBN 978-1-61519-359-2

Cover design by Sarah Smith
Author photo by Ben Queenborough
Manufactured in the United States of America
Distributed by Workman Publishing Company, Inc.
Distributed simultaneously in Canada by the University of Toronto Press
First printing August 2016
10 9 8 7 6 5 4 3

# Contents

good. Sam was simply indulging my rookie pride in my still-fresh knowledge by testing me. He knew I had recently passed the exam and test to become a Yachtmaster. Perhaps he had fond memories of that stepping stone in his own career.

Sam regaled me with something scarier than tales of life at sea. Nothing he had seen at sea was more terrifying than facing the Board for the oral exam to become a Master at the Warsash Maritime Academy. Sam's vicarious pleasure in recounting the ludicrous level of detail that needed conquering in this professional rite of passage was obvious. "They might allow you a mistake, but probably not two. And if they smell any weakness in your knowledge they are merciless . . . Predators!"

The nautical rite of passage was beautiful in itself to me. Qualifications soften the self-doubt any honest twenty-something will admit to feeling. If somebody gives you a piece of paper and says that you have passed an exam, then they know best, maybe you do know something. And if you know something, then maybe you are worth something.

ALTHOUGH I SHOULD HAVE reveled fully in that first professional voyage, there remained an awkward doubt. Even with that piece of paper, complete with identification photograph, slipped into a neat Royal Yachting Association wallet. There was still a worry that gnawed at me and chafed my mind like old hemp rope running through my hands. This anxiety took the form of Captain Abharah.

Everywhere I looked I saw Captain Abharah. It did not matter which salted guardrail I peered over or which slopping slice of gray North Sea I looked down into, there he was. He even retired with me when my watch was over, joining me in my rolling bunk. He was a disconcertingly persistent companion and the small fact that he died a thousand years before I was born proved of little consolation to my troubled mind.

Captain Abharah began his working life as a shepherd in the Persian district of Kirman. A job on a fishing boat took him to sea, and this led to work as a sailor on one of the vessels that traded the route to India, before he hopped to plying the treacherous China sea lanes. At the time, it was felt that nobody ever made the journey to China and back without serious mishap. Abharah did it seven times, and all of this took place toward the end of the first millennium.

How do we know this much about a man of lowly background, from a distant part of the world, all those years ago? Because he did something that displayed extraordinary knowledge and chutzpah. Enough of each for his story to survive.

Once, a sailor who also worked the feared route to China, Captain Shahriyari, was waiting out a worrying calm in the midst of the typhoon season, when he spotted a dark object on the distant sea. A small boat was lowered and four sailors were dispatched to investigate this mysterious speck. On reaching the dark object, they discovered a familiar face; the respected Captain Abharah was sitting calmly in a canoe, with nothing but an animal skin full of water.

When they returned and reported this surreal sight to Shahriyari, he asked why this stricken and respected other captain had not been rescued and brought on board. The crew replied that they had tried, but that Captain Abharah had refused to swap his small canoe for their greater vessel, claiming he would be fine on his own and would only join them if he was paid the substantial sum of 1,000 dinars.

Captain Shahriyari and his crew mulled over this bizarre proposal, but with their estimate of Abharah's wisdom and their fear of the strange weather conditions that prevailed—the lull obviously causing them concern about what lay ahead—they agreed and lifted Abharah on board. Once on board his new ship, the canoeing captain wasted no time in demanding his 1,000 dinars and he was duly paid. He then told Captain Shahriyari and his crew to sit down, listen, and obey his orders. They did.

"*Al-daqal al-akbar!*" Captain Abharah shouted.

Abharah explained to the captain and his crew that they were in great danger: They must throw their heavy cargo overboard, saw down their main mast and throw that overboard, too. Then they were to cut the cable for their main anchor and let the ship drift. The crew obeyed Abharah's orders and set to work, although it can't have been easy—three things a sailing trader would prize above most others were his cargo, the mast, and the main anchor. These were their tangible icons of wealth, transport, and safety—the reason they had risked their lives and their means of preserving them. But they did as they were told and then they waited.

On the third day a cloud rose up until it appeared like a lighthouse before them, before it dissolved and crumpled back down into the sea again. And then the typhoon—*al-khabb*—hit them. It raged for three days and nights. The new lightness of the ship allowed it to bob over waves and reefs like a cork and they were saved instead of swamped, smashed, and drowned.

On the fourth day the wind died down and the crew were able to make their way safely to their destination in China.

On the way back from China, the ship now freshly loaded with a new cargo, Captain Abharah ordered the vessel to stop. The dinghy was lowered and some sailors were sent to find and retrieve the great anchor they had cut loose before the storm and left on the reef.

The crew were stunned and asked Captain Abharah how he had known where to look for the anchor and how he had forecast the typhoon so accurately. He explained that knowledge of the moon, the tides, the winds, and the signs in the water made all of it perfectly straightforward.

AND SO IT WAS that Captain Abharah's deep intuition and understanding haunted me on that voyage from Norway. The wisdom that allowed Abharah to read the signs was not to be found in any of the exams I'd passed, but it most certainly existed. Traditional Arab navigators had a word for this body of knowledge, one that allows someone to read the physical signs in the water—the few with this ability possess the *isharat*.

Obviously, I thought, this wisdom comes from a different source than official exams: It is to be found in time at sea. And so I went about building this wisdom by spending days, nights, weeks, and months at sea.

But I was wrong. Time at sea on a modern yacht teaches you how to manage a boat and a crew, how to read between the lines of a synoptic weather forecast, how to make bread in a rolling galley, and how to enjoy raw fish with the help of a little lime juice. It does so much, but in an age of dazzlingly capable electronics, it doesn't address the stare of Abharah. It no longer offers that deep wisdom, it no longer teaches us how to read water. I have often discussed this with modern captains of great experience and they universally agree, often with a sadness in their horizon-drawn eyes.

DELIGHTED BY THE EXPERIENCE of time on the water and equally frustrated by the lack of wisdom it was giving me in my ability to decipher the patterns of the water I found around me, I changed course. Many years ago I embarked on a parallel voyage, this time in search of this wisdom. And as soon as I set out on this journey, something very strange happened. I quickly discovered that the clues that unlock a deeper understanding about the water around us are not found in greater proportion the further from land we are. The things we notice in puddles and streams can be just as profound and helpful to understanding what is happening as those that might be spotted from a vessel mid-Atlantic.

Secondly, a related discovery: It is actually easier to learn about water with your feet on firm ground than in a boat, whether or not you intend to use this knowledge on a boat later. And so in this book, I will, wherever possible, illustrate how these things can be not just learned on dry land, but witnessed and enjoyed from there, too. This may sound far from ideal or even far-fetched, but it turns out that it is a tried and tested approach, used by some of the greatest readers of water signs humanity has ever produced.

The Pacific Island navigators have been stunning Westerners for centuries. Captain Cook encountered these formidable seafarers in Tahiti in 1774, when he watched as 330 vessels and 7,760 men took to the water. Cook and his companions were "perfectly lost in admiration."

Without the use of charts, compass, or sextant, the Pacific Islanders found their way over great stretches of ocean, relying entirely on their interpretation of nature's signposts. The Islanders' reading of water in particular has never been bettered by humans, anywhere on Earth. We will be getting to know their methods in the coming chapters, but the reason for introducing them here is to discuss their methods of passing on their unique level of skill to the next generation.

Just as there is an Arabic word for the body of knowledge about water signs, so there is an expression in the Pacific: *kapesani lemetau*—the talk of the sea, the water lore. The young Pacific Island students of this wisdom would go sailing with their tutors, but the finer parts of the art were actually passed on while still on land. Many lessons about stars, wind, and waves would be taught indoors. Teeta Tatua, a *tia borau*, or navigator, of the Gilbert and Kiribati Islands in the Pacific, was taught his skills by his grandfather in the *maneaba*—the meeting house. Many others were taught their skills using a "stone island" or "stone canoe." This is a simple teaching aid used to demonstrate to a student how the water would behave around them and how to interpret it, from the comfort of sitting on a beach.

We should be inspired by the Pacific Islanders to appreciate both what is possible and how we can learn so much while still on dry land. But we must not be intimidated by their abilities. In the dated words of a legendary Australian bushman and

to order. If you find a sign in this book that you like
and look for it, you may well find it on your first attempt,
more likely that it will appear before you at a time of its
g—provided you retain enough curiosity to keep look-
it. This means that the best approach is to view this art
ole, and this book is structured to allow you to embark
est to get to know all the signs, while remaining aware
ch one is part of a bigger jigsaw. This will make you
not only seek individual signs, but to meet water in all
ds and in whatever guise it appears.

re will be challenges, frustrations, and possibly even a
nfusion when you meet some of the more complex pat-
the first time. I encourage you to think of the signs and
will meet as "characters": some are straightforward,
more complex ones are often the most interesting ones
e.

ly, you might reasonably be asking why you would want
the effort to take on this rare quest. I will let Chad
Baybayan, a modern-day *Pwo*—master navigator—in
fic answer that. When Chad was interviewed in 2014
d whether there was any point in studying these meth-
e modern world, he replied:

ally is a pretty unique set of skills that one would
re to becoming proficient at. What it truly does
arpen the human mind, intellect, and ability to
pher codes in the environment . . . For me, it's the
euphoric feeling that I have ever felt.

conservationist, Harold Lindsay: "D[…]
natives possess powers that are denied[…]

Not only are we able to emulate t[…]
we can combine this insight with mor[…]
experience, and wisdom. Ian Proctor, [a…]
strategist, who has helped racing crew[…]
in the world, declared that many sail[…]
anyone gets in a boat. How? By readi[ng…]

IN THE FOLLOWING PAGES I have[…]
of water behavior that I think are w[…]
long list gathered over time, I have se[…]
are the gems that I think encapsulate[…]
useful. However, for me to give you t[…]
this art, there are two obstacles we [must…]

The first is the way that natural hi[story…]
up into its realms: ponds, rivers, lake[s…]
to be very different to each other. If[…]
the animals and plants, then this is[…]
very few creatures or plants will be[…]
lake and the sea, even if they are onl[y…]
However, the water itself does not h[…]
boundaries, and we can learn a lot a[…]
world's greatest ocean by looking at[…]
your favorite type of water, the th[…]
not and cannot be confined to tha[t…]

Secondly, the study of water si[…]
fectly to an impatient check-box[…]

perf[…]
and[…]
but [i…]
choc[…]
ing f[…]
as a […]
on a[…]
that [.…]
ready[…]
its m[…]

Th[…]
little c[…]
terns f[…]
clues [.…]
but th[e…]
over ti[…]

Fina[…]
to mak[…]
Kālepa[…]
the Pac[…]
and ask[…]
ods in [.…]

It [r…]
asp[…]
is s[.…]
dec[…]
mo[…]

The Pacific Islanders attach huge importance to the process of learning these skills. Their induction into this elite world of rare knowledge and the accompanying initiation are accompanied by traditional rites. The details of this training and inauguration vary from island to island, but there are some common themes. A special loincloth is worn, and the initiated will be dusted in turmeric and exchange gifts with friends and family. During the whole process, which may last as long as six months, they will be expected to remain celibate and drink special coconut potions, while abstaining from water. With my love of wisdom-gaining rites of passage, you can probably imagine how much I adore that.

You will be able to choose your own way to celebrate learning how to read water. But if you ever see water in exactly the same way after reading this book, then I will have failed in my own task and there will be no coconut potions for me.

I hope you enjoy the quest.

*Tristan*

# 1

# Launching

OUR JOURNEY WILL BEGIN, like those of so many great explorers before us, in the kitchen.

One of our few expectations when we look at water is that the water should lie flat, but water is rarely flat. Look very closely at a glass of water and you will notice how the surface of the water in the glass is not flat; it curves up slightly at the edges. It has a "meniscus." This meniscus curve is caused by the water being attracted to the glass. It is being pulled by the glass and then sticking to the edges. The attraction between the water and the glass turns what would otherwise be a flat surface into the gentlest of bowls with a tiny rim.

What is the use of us noticing that? On its own, perhaps not very much. But by drawing a few pieces together it can become a stepping-stone to helping us understand why a river will flood.

The water is attracted to glass and this is a feature of water. A few liquids, like the only liquid metal, mercury, are repelled by glass, and this leads to an upside-down bowl shape or "convex meniscus." Most liquids will show either an attraction or repulsion for other substances. Liquids are also weakly attracted to themselves—if they were not they would separate and become gaseous. Water is attracted to water.

Water molecules, as our science teachers drummed into us, have two hydrogen atoms and one oxygen atom, and these are bound together tightly. But the thing that the teachers—mine at least—don't tell us is that hydrogen atoms in one water molecule are also attracted to the oxygen atoms in the other water molecules nearby. And this makes water stick to water. It might help to think of two balloons that have been rubbed on a sweatshirt that then stick weakly together due to the static. The science is similar, albeit on a minute scale.

It's very easy to demonstrate this stickiness of water. Take a glass of water and pour a few drops onto a flat, smooth, waterproof surface, like a kitchen countertop. Now sink down until your eyes are level with the droplets. Do you see how the water forms itself into a series of gently bulging little pools? It doesn't sink until it is totally flat and then all run off the tabletop (some will run off, if you pour enough, but some will also remain). Instead of flattening entirely and running off, there is now a group of small upside-down puddles.

This happens because the water's attraction to the other water nearby, its stickiness or tension, is strong enough to resist gravity's pull. Gravity is trying to pull the water down, so that it flattens and then runs onto the floor, but water's tension is strong enough to resist this. It is one of the reasons we are more likely to reach for a cloth than a mop when someone knocks a glass of water over. The water still on the table pulls the water back and stops it all from running onto the floor.

Pick two of the bigger pools that are reasonably close to each other. If you put your finger in one and draw it toward the other and let go, not much happens—the pool may stretch in size a little, but that's all. Notice how it has a tendency to shrink back a little bit, as the water you have pulled with your finger gets gently yanked back by the attraction of the water it has left behind. (If you try this on different surfaces, you'll notice how the amount and rate that the water shrinks back varies from one surface to the other, because it depends on how much each different surface attracts the water.) But now if you draw your finger further, all the way until the two pools just touch each other, watch what happens. The water that was being pulled back by its home pool no longer is. Instead it is now being pulled by its new friend—the pools join together as one, bonded by water's stickiness.

After one of these experiments, when I came to clear up and pulled a cloth across the small pools, the water did something that it always does, but which I had never previously noticed. The cloth absorbs a lot of the water—that's its job—but the remaining water is "ironed" out into a thin flat layer. But this layer only stays flat and thin for a second, before the water quite literally pulls itself together, forming hundreds of very small pools again. These tiny pools are often interlinked, giving a mottled appearance to the wet area. Try it and you'll see what I mean.

LEONARDO DA VINCI was fascinated by water and carefully observed its "stickiness." He liked to watch the way a small drop of water does not always instantly fall from the underside tree branches. Da Vinci noticed that when the drop is big enough to fall, it does so with some resistance. Around 1508 he noted the way that before a drop finally falls, it stretches until a neck of water is formed, and when that is too thin to support the weight of the drop, only then does it fall.

You can spot this effect for yourself. It appears rather beautifully at the ends of leaves after rain. If it is still raining heavily, the water will flow in spurts off branches, twigs, and leaves, but shortly after it stops raining, take a look at the leaf tips of a broad-leaved tree or shrub. The water collects and often runs down the thin rib at the center of the leaf, before gathering at the tip. The drop hangs there, the tension or stickiness of the water is now battling gravity, before enough water gathers, gravity prevails, and the drop falls. The leaf often bounces up elegantly at this point, and then the process begins again.

The place that this tension in water is most apparent is at the water's surface. Since the water molecules near the surface are being pulled downward by the molecules below it, but not being pulled up by any molecules, this leads to the surface coming under tension, which in turn gives water a sort of thin skin. There is a simple experiment you can do that proves two fundamental things: that water has a skin formed by surface tension and that this tension is the result of the weak bonds between water molecules.

For this trick, by which I mean, serious experiment, we are going to prove that the water's surface tension creates a skin that is strong enough to support a small metal weight. To do this we are going to watch a needle float on water. The only challenging part is the first bit, because we need to rest the needle on the water very, very slowly and carefully, otherwise the needle will break through the surface of the water and sink to the bottom. There is a sneaky way to do this: Rest the needle on a small piece of blotting paper (a little harder to find these days, but still in most stationers). The blotting paper will slowly become saturated and then sink to the bottom of the bowl of water, leaving the needle floating.

This proves that the surface tension of the water is strong enough to support a small piece of metal. Now we need to prove that it is the electric bond between the molecules that creates this skin. We can weaken the bonds between the water molecules by adding a little detergent to the water. Any dish soap will do—detergents work partly because they carry charges that nullify the electric attraction of the water. The needle sinks.

If you approach a large body of still water—a pond or lake—anywhere outdoors near summer, you are likely to find a busy world of insects. And by observing the insects you will see the water skin experiment at large. Head toward the sun and keep down low if you want the best effects; these insects are very sensitive to things that swoop in on them, so your best chance of catching them unaware is if you move slowly and stealthily toward the light. On a sunny day, if your shadow is directly behind you as you reach the water, you'll see a lot more insects.

There will be insects in the air and many underwater, but some of the most interesting ones are the ones sitting on the surface. Why don't they fall in? We certainly would. It's because the surface tension of the water is stronger than the effect of gravity on small insects. For big lumps, like humans, it is the other way around, but this does at least make swimming more enjoyable. Don't worry too much about what these insects are at this stage. We will be getting to know a few in time, but it is worth admiring how nature has evolved to make the most of the water's surface tension. This is one of the many, many reasons why detergents and wild water do not make a happy combination.

The same tension that leads to water sticking to itself and the sides of glasses is also responsible for something called "capillary action." We're all at least a bit familiar with the idea that liquids don't always obey gravity. Every time we dip a paintbrush in water we watch the water flowing upward into the hairs, even though our understanding of gravity tells us that water shouldn't flow upward in this way.

The reason for this capillary action is a simple combination of the two effects we have been looking at. Water is attracted to some surfaces, like glass and paintbrush fibers, and it is also attracted to itself. So when an opening is thin enough, something interesting happens: The meniscus effect means that the surface of the water is attracted to the material above it and is drawn upward, and since it is a narrow opening this pulls the whole surface of the liquid upward. Then, because the water is sticking to itself, the water just below the surface also gets pulled along and follows it upward. The narrower the opening, up to a point, the more dramatic the effect.

Every plant you see, from a tiny weed to a great oak, depends upon capillary action to get the water from the ground to its highest leaves. We know there are no pumps in trees and yet thousands of gallons—tons of water—have to get from the soil to the tops of tall trees somehow. Without capillary action this would be impossible.

Back in the kitchen, the reason that reusable cleaning wipes, tissues, and other finely woven materials are so good at mopping up water is that they have been specially designed to maximize capillary action. There is something strangely satisfying about the way a really good cloth will mop up the water around it, like a magnet, without you needing to move it. That is the satisfaction of capillary action.

It is time to look at this effect in a wilder context. The next time you pass a small river, stream, or ditch that has muddy banks, take a look at the mud of the bank. We would expect the mud to be dark and wet where the water is splashing it, but notice how it appears wet higher than the water splashes, higher than any water appears able to reach.

The mud above the water is a mixture of particles and air gaps, a little like a fine honeycomb of thin tubes. The water gets drawn up into these gaps by capillary action with the result that the mud becomes saturated above the level of the water in the ditch or stream. The distance that water can travel upward is influenced by a number of factors, including its purity—clean water rises higher than polluted water—but the main one is the size of the gaps between the particles. Water rises much higher in soils with fine rounded particles, like silts, than in course soils, like sandy ones. At the extremes, water can rise very high in clay, but will hardly rise at all in gravel.

The air pressure will also affect the amount of water that rises up through the soil and is then held there in suspension. This means that when there is a sudden lowering of air pressure, as we get when storms are approaching, the soil is unable to hold onto as much of this capillary water and it drains out very quickly into the local streams, adding to the likelihood of flooding during the storm.

IT IS WORTH taking a short detour at this point to demonstrate how noticing the smallest of things can combine with broader observations to give us a deeper insight into what is going on. Let's look at how a messy kitchen experiment can combine with a walk on the beach to help predict whether a local river might flood.

The height of the sea is influenced by the state of the tide, which in turn is influenced by many things that I will be covering, but here I'll just mention that these include the atmospheric pressure. When the air pressure is low the sea will be higher than when the air pressure is high—a difference of about a foot—is typical from a big high to deep low pressure system. To help you remember this, think of a high pressure system and its lovely blue skies pressing down on the horizon, lowering the sea.

Imagine you're in a coastal area you know well and you suddenly notice that the sea appears higher than you have ever noticed it before, even at high tide. That might lead you to suspect that the air pressure has dropped quite a lot. This in turn means that you can predict not just bad weather approaching, which is likely when the barometer drops, but also an increased risk of flooding, as some of the water being held by capillary action above all those streams, ditches, and rivers is about to be released, before the first raindrop has even fallen.

Once we know the things to look for and what these influence, every patch of water we see is beautiful, fascinating, and a clue to something else. We learn to see the water as part of an intricate network, a matrix if you like. At various times these skills have been called magic and more recently, psychic—they are neither. They are the fruits of a little curiosity, awareness, and a willingness to join the dots.

In this whirling vortex of a chapter we have looked at water in the kitchen, on leaves, in streams, and by the sea. On our journey to emulate the greats, like Superaga, the fourth-century Indian water expert, who "had a deep knowledge of the value of the signs," we must get used to the idea that an understanding of water in one area will help in another.

# 2

# How to See the Pacific in a Pond

DESPITE OUR FREQUENT TRIPS to the coast to swim in the sea, my family's craving for water and swimming got the better of us and a few years ago we began scheming. In chalk country there is not much standing water, as it tends to percolate downward, and the options for wild swimming in ponds are seriously limited. It appeared to us to be a clear case of Muhammad and the mountain. If we couldn't find a wild pond to swim in, then . . . There is now a not-insubstantial pond in our garden, which we swim in most of the year.

There is a long list of gardening jobs that I loathe; maintaining our pond, however, remains a joy. There are always a few things to be done on any given weekend: brushing, netting, skimming, cutting back the water plants, tackling over-enthusiastic algae. Strangely, I never seem to tire of these jobs. The result of this fun, combined with my love of and fascination with water, is that I have spent an extraordinary amount of time pond-watching. Only this morning I counted fourteen frogs and thrilled at the sight of the oozing black frogspawn filling the gaps at the base of the crew-cut prespring plants.

Last year I was heading out to meet someone when I paused at the pond's edge and looked in, as I invariably do, even when running late for things. Then I tried to leave the scene—as a policeman would put it. But I couldn't. The usual magnetism I feel for water was even stronger than normal. Glancing at my watch, the small sensible part of my brain nagged the larger, irreverent part to move on, but there was something in the water that would not let me go. Then I saw it, or I should say, then I became aware of what it was, the two things not being the same at all.

Our brains are contending with so much information from our senses that they rely on a filter to cope. There is an automatic prioritizing system in the software of our heads and it constantly sifts the information that our eyes relay for things of urgent interest. In evolutionary terms, we would once have been most interested in predators and prey—threats and opportunities. And both predators and prey move, which is why we notice movement in any scene, before spotting more subtle clues. Everybody notices the rabbit run across the path, but very few spot the pile of leaves to one side of the path; until, that is, the wind catches the leaves and creates motion by blowing them across the path.

When we look at any body of water, this same filter is at work. We will spot motion in the water before we see any subtle shifts in color or shade. The wind was quite strong that day and was blowing across the surface of the pond. At one edge of the small pond, there are some half-submerged rocks that we use as stepping-stones. My eyes had been drawn to the ripples that the wind was creating in the surface of the pond's water, but it was not this simple effect that we have all seen a thousand times that was captivating my attention. What I was looking at, and struggling to decipher, was that the patterns in the water around the stones were resonating with knowledge of how water behaves in a very different part of the world.

IN 1773 CAPTAIN COOK was at his most alert as he sailed close to a treacherous area of the Pacific, called the Tuamoto Archipelago. The islands had been nicknamed the "Dangerous Archipelago" by sailors who had known too many ships smashed on their reefs. Cook couldn't see the islands or the scattered reefs that surrounded them, but he knew they were there because he could feel them. Cook didn't have a psychic sixth sense; he was tuned to the behavior of the water and had noticed that the swell, which should have been running from the south, causing waves that should normally have been easy to feel, was definitely absent. It was straightforward for him to deduce that the archipelago must have been to his south and was shielding him from those waves. The water was calmer, Cook realized, because he was in a "swell shadow." As soon as he felt these waves return, Cook was able to relax a little, knowing that he must have passed the area of danger.

Looking at the stepping-stone in the pond I could see that the ripples were reaching it in sets, as the wind blew over the water. But just downwind of the stone there was an area of calm water. It was the only area of calm water anywhere near the center of the pond. This was a "ripple shadow," an area protected from the wind-driven ripples by the stepping-stone, and it reminded me of the swell shadow that Cook had sensed.

Although Cook was an extraordinary sailor and navigator, he was only familiar with the more basic water-reading techniques and ignorant of many of the more elaborate ones that were already well established in the Pacific at this time. We are all able to understand more about these intricate signs now, thanks to academic investigations of the past century, than Cook ever did. It was some of these more refined and beautiful patterns that I first saw in the pond in front of me that made me both late and happy. After that first time, I have noticed these patterns in ponds, lakes, rivers, and seas near home and far away. These are signs that we can all find if we choose to look for them.

THERE WERE FIVE clear patterns in the water around one of the stones. There was the "open water," the main part of the pond, where the wind sent ripples across the surface in an orderly way. Then there was the "ripple shadow" on the side of the stone that the ripples could not reach and where the water was calm. Three further patterns were also recognizable in the water.

As the ripples hit the stepping-stone, some of that energy bounced back, like an echo. This meant that on the side of the stone that the ripples were arriving from—the side opposite the "ripple shadow"—there was an area of choppy water, and this chop was being caused by the arriving ripples running into the ripples that were being reflected back. In this small area, the water was behaving in a way that was different from anywhere else in the pond. Looking to either side of the stone, I noticed that there were two patches of water that behaved in a similar way to each other, but in a way that was also different from the rest of the pond. Finally, there was a line where the ripples met once more on the far side of the stone, meshing back together and creating their own pattern.

I suddenly appreciated that I was looking at a "ripple map": the ripple patterns were related to the location of the stone, according to strict rules and laws of physics. These ripple maps corresponded very well with the swell maps that Pacific Island navigators have used for centuries to find their destination island, a vital skill when searching for a small dot in a great ocean. Before my eyes the stepping-stone in the pond metamorphosed into an island in the Pacific.

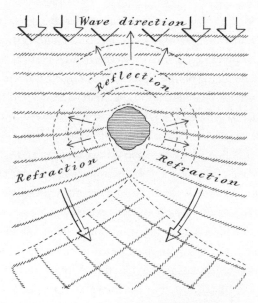

*The ripples around a stone in a pond resemble the waves around an island in the sea.*

Here it is worth introducing the idea that there is a difference between ripples, waves, and swell. All three are water waves created by the wind blowing over the water. Ripples are created almost instantly and die away as quickly when the wind dies down. You can make ripples by blowing into a cup of tea. Waves require wind to blow over a bigger area and will not die down straight away if the wind stops, but will within hours. Swell is the name of the waves that have enough energy to travel beyond the area of wind. We will look at these different types of wave in more detail in the Reading Waves chapter, but for now we can think of ripples in a pond as being the same as waves in the ocean.

As I LOOKED DOWN onto the pond's surface, growing ever later for something possibly more important and certainly less beautiful, I imagined being one of the small, dry, broken leaves that were bouncing on the ripples. That bounce changed as the windblown leaf passed around the stone, and if I had been an ant on that rocky leaf, I would have had a chance to feel where I was relative to the stone island. This is the art that the Pacific navigators called *meaify*, the fine skill of navigating by reading the behavior of the water. The motion is sometimes easier to sense with your eyes closed, and some navigators have been known to lie on the deck with their eyes shut.

We know more about the way some Pacific Islanders read the patterns in the water than much of the rest of their culture, thanks to the curiosity of Captain Winkler of the German Navy in the 1890s. Aided by his interpreter, Joachim de Brum (who went on to become a master navigator himself), Winkler's studies in the Marshall Islands preserved a sublime and unique example of water wisdom.

The Marshall Islands sit near the equator in the Pacific and are part of the Micronesian group. Lacking mountains, these islands lie low in the sea and cannot be seen by seafarers until they are close. In a nautical world, one where life depended on traveling by sea and finding islands, but one without compasses, charts, or sextants, the conditions were perfectly set for a rich and sophisticated culture of water-reading to blossom.

Captain Winkler discovered that the Marshallese saw the water in much the way that a European cartographer might view the land: not as a melee of water that shifted with every change in the weather, but as a terrain laid out with a series of recognizable features. The depth of the sea has always been of huge importance, and the nature of the seabed was sometimes recorded, as it helped with both navigation and picking spots to drop anchor, but the idea that the character of the surface of the open ocean was worth mapping was alien outside the Pacific. This was the received wisdom of European sailors for most of our history. About the only exception to this view is that water conditions change close to land. The seas may become confused and choppy, but by that point you're usually in sight of land anyway, so the water's surface was generally considered of little importance or relevance in terms of understanding where you are for longer-distance navigation.

Deprived of any alternative approach, the Marshallese took the opposite view. Once on land the navigation was over for them, their challenges lay between the islands, at sea, so they learned to view the sea much more forensically.

The islanders observed that the wind comes from certain directions fairly dependably—these are the prevailing wind directions, and everywhere on Earth has its own trends. These prevailing winds set up predictable sets of swell in the ocean, and when these waves hit an island, equally dependable things happen. On each side of an island, the water would behave in a telltale way. The waves that hit the island head-on would bounce back and mix with the oncoming swell. The waves that passed close to the island would get bent and set up a different pattern on either side of the island, and on the far side of the island, there would be a swell shadow.

The genius in these skills lay in two simple and connected observations. Firstly, winds are seasonal, so the waves they create are broadly predictable and the patterns that will be created around the islands will also be predictable. Secondly, these patterns can be used to deduce where land is from the behavior of the water. Just as a land navigator can deduce the direction a river is likely to be found from the gentle downslope of a hill, so the Pacific Islanders could deduce the direction of an island from the particular rocking motion of their boat. This knowledge and the accompanying skills have been found to be widespread across the Pacific Islands. Each island community may have its own set of local patterns to interpret, learn, and pass on, but academics have found the similarities between even distant island groups to be greater than the differences. This shouldn't surprise us: They faced similar conditions, had similar needs, all lacked even basic navigation technology, and their wisdom spread as there was much cultural exchange between the islands. Above all, water will obey the same laws around one island as it does around another, even if it is many miles away—even if it is a very different size. Even, as I discovered, if the island consists of a stepping-stone in an English pond.

IN THE MARSHALL ISLANDS, Winkler found something there that is unique in human history. It was on these islands that Winkler or anyone else has found a physical object that represented this water wisdom. The Marshallese navigators made "stick charts," using palm ribs that were bound with sennit (a type of cordage made from dried fibers), to fashion something that represented the various interlocking swell patterns that the sailors should expect to find out at sea. The "stick charts" were not charts in the Western sense; they were never taken to sea or used to represent the real world exactly. Instead, they were teaching aids, used by an experienced Marshallese navigator to teach an initiate.

IF IT WAS Captain Winkler who lit the fuse of Western interest in Pacific methods, then it was the English-born, New Zealand–reared sailor and scholar David Lewis who created the bang. Lewis spent long periods in the 1970s sailing with and interviewing the Islanders and did more than anyone else to spark a renaissance of Western interest in this field.

The wisdom held within the Marshallese stick charts had not died out altogether, and David Lewis sailed with the last remaining navigators to retain this knowledge. Lewis joined local *tia borau*, or navigator, Iotiebata Ata, on a short voyage from the island of Tarawa to the neighboring Maiana. They set sail in Iotiebata's thirty-foot canoe, a sleek, speedy vessel that he used for racing and shark-fishing, for a voyage of only eighteen miles, but the distance was not important. Along the way Lewis was able to watch as Iotiebata pointed out where they were relative to each island, explaining it all with reference to the behavior of the waves.

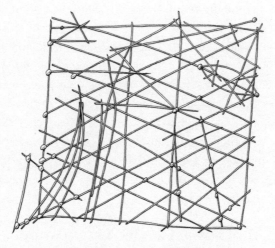

*A stick chart from the Marshall Islands.*

Iotiebata demonstrated how the water changed as the easterly swell was bent by each of the islands, and as he did so, the blue melee of the sea was being transformed into a chart before Lewis's eyes. Iotiebata was also able to point out where smaller waves were riding on top of bigger ones, superimposing their pattern on the bigger, more dominant swell. By reading through the temporary waves to the underlying swell and detecting the way it had been influenced by the distant islands, Iotiebata was able to sense and "map" the location of invisible land.

On another research trip, Lewis described the navigator, Hipour, as able to recognize familiar swell patterns "like people's faces." Some were familiar enough to become friends and earn their own names; one very familiar one was known simply as the Big Wave, and it had a special significance as it came from "under the Big Bird." In Pacific Island navigation, direction was not referred to in terms of north, south, east, or west; instead, they used the names of the stars that would rise or set in that direction. The Big Bird was the local name for Altair, a star that rises in the east. So, by describing a swell as the Big Wave from under the Big Bird, they were able to identify both the character of the swell and its direction.

It is worth considering how extraordinary these skills are. Every sailor of a little experience learns to sense the difference between various sea states, and there are enough anecdotal stories of skippers who have deduced something about their location from a barely perceptible change in the ocean's motion. There is also the eighteenth-century legend of Captain Edmund Fanning, who reportedly woke up one night, rushed up onto the deck and ordered the crew to heave-to, the equivalent of slamming on the breaks in a sailing vessel. It was not until the following morning that Fanning and his crew appreciated that there was a reef that would have wrecked them less than a mile away. Fanning had felt the reef, through the change of the water's behavior, in his sleep!

But this is a legend, with vague sources, for a reason. It is very unusual for Western sailors to be able to read the sea in this way. I have tried to practice this skill on many occasions, and it has only heightened my respect for the Pacific navigators.

These are fine skills and, in truth, hard for any of us to develop to the levels of the Pacific Islanders without giving up much of our lives to that pursuit. However, we should not be surprised by this and must not be dispirited—it may be hard to use these skills to cross the Pacific, but these are patterns that we can all still see and recognize when we look at water much closer to home. It continues to amaze me that I now see the way the ripples reflect and bend around the stones in our pond at home, despite never having noticed them for hundreds of hours of pond-gazing before. As I mentioned earlier, there is a difference between what we see and what we are aware of.

WE WILL BE RETURNING to this area in the Coast and Reading Waves chapters, to delve a little deeper and look at a multitude of other signs and patterns, but before we leave the pond, I'd like you to seek out a pair of paddling ducks on a relatively calm pond.

Ducks are companionable creatures; you will very often see the female accompanied by her more colorful drake. (Drakes get the fancy colors, the bottle-green head, white necklace ring, and bright yellow bill, but females get stuck with the plainer lines. This is because the female needs better camouflage for the very vulnerable time she spends sitting on her eggs.)

Take a good look at what the water is doing all around your ducks. It may be placid if it is a calm day and there are no other birds churning things up. It is more likely that there will be a mix of ripples from a breeze and some disturbances from other birds. Whatever you see, just spend a moment getting to know the shapes and rhythms of the water, not just by the ducks, but over the area you are looking at more broadly. This is the "baseline water behavior." In other words, this is what the water is doing before it is disturbed by our chosen pair of ducks.

Now look at the water behind one of the ducks as it paddles through the water. You will quickly spot the V-shaped wake behind it, a series of ripples that have been created by this one bird and that spread out and over the water, superimposing themselves on whatever patterns were there before. Next, look at the very similar ripples behind your duck's companion and watch those march out across and over the water behind them as they both swim.

*Duck wakes create a new pattern where they meet.*

Examine the place where these two patterns touch and then overlap each other and study this for a few seconds. Can you see how a totally new pattern is created, one that is formed by the combination of the sets of ripples generated by each duck, but which looks different from each one individually? You should be able to see a new criss-cross pattern.

The water's behavior where the two sets of waves overlap is unique, which is why the Pacific Island navigators were able to work out where they were when two sets of swell ran into each other near islands. The waves bouncing back off an island meeting the oncoming ones create one pattern, for example, as do the waves wrapping around one side of an island meeting those that were bent around the other side, meeting eventually on the far side of the swell shadow and creating another pattern beyond the swell shadow.

When waves run over each other like this and create a new pattern, scientists call it "wave interference." Where two crests coincide, the water becomes double the height it was before, and where two troughs coincide a doubly deep trough is created, but where a crest from one set of ripples or waves meets the trough from another, they cancel each other out. The result is a patch of water that is born of two sets of waves, but looks different from either of them. We will be seeing this important effect in different places and on different scales later in the book, but the ducks and the Pacific Islanders are not a bad team to introduce us to it.

# 3

# Land Ripples

IN 1885 the South Australian Government dispatched a man named David Lindsay and a small team of surveyors from Adelaide into the arid country of the Barkley table-land. By February of the following year they were still at work when they were confronted by the desert traveler's nemesis, an enemy known to the local aborigines as *quatcha queandaritchika*. It can be loosely translated with the less delightful-sounding English words, "water all gone."

The Todd riverbed was dry, and all around the land was parched and dusty, so the party was down to its last few pints of water. They were now in a serious survival situation. One of the party headed out on a camel to search for water, but returned later, exhausted and totally dispirited, having found none. Worse than that, he had seen no signs of the Aborigines themselves. The surveyors had long appreciated that in this most testing of environments, the locals' relationship with the landscape was a clue as to how hospitable or otherwise it might be. Signs of fires meant the Aboriginals had made camp, which invariably meant that water would be found very close by. But there were no signs of camp having been made for miles around. Things looked bleak.

Fortunately there is one clue to the location of nearby water that is even more dependable than the presence of the Aboriginals. The explorers were beginning to feel the first dangerous pangs of hopelessness when Lindsay spotted a lone, wild rock pigeon flying across the gorge. Immediately appreciating the angelic significance of the bird, Lindsay set off in pursuit. The bird may have disappeared out of sight, but he had noted its flight path and tracked it up a hill. In an unlikely spot, one where he would never have considered looking, he found a hole in the rocks and beneath it more water than he and his party could drink in a year.

The likelihood that any of us will find ourselves desperately short of water in the outback or anywhere else may be slim, but the opportunity for us to read where the water is, relative to us, is an important part of a water reader's armory. Our view of our surroundings can once more be inspired by those extraordinary navigators in the Pacific, and they can offer us a perspective that can be very helpful anywhere in the world, from a city center to the wilderness. In this chapter we will focus on how it is possible to see the water before you see the water.

PACIFIC NAVIGATORS do not aim precisely for their destination island; instead, they head as best they can toward the area of ocean that they know the island is in. Once the navigator judges—from the length of passage and other signs like the position of the stars—that the island cannot be too far away, he begins to scour the sea and sky for the clues that help to make invisible land appear. As well as the swell patterns we looked at earlier, one of the main clues was the species of birds that could be seen, as each species could be used to estimate distance from land. Terns, boobies, and frigate birds each have their own comfortable range from land, and so a flock of any of these formed part of the navigator's radar. Frigates might roam as much as seventy miles from land, whereas terns were a very welcome sight, as they were rarely seen more than twenty miles from land and so meant that landfall could be expected soon. This use of birds to work out the distance to land is such a fundamental part of the natural navigator's trade that it has appeared in stories as diverse as the Old Testament, where Noah released a dove to gauge whether the water had receded, to Norse lore. In the Pacific, this technique has earned its own name: *etakidimaan.*

The sea life also changed noticeably as land drew nearer, as fish, dolphins, and jellyfish, just like all other animals, have preferred habitats that will be strongly influenced by the shallowing of the sea nearer land. But there were other signs, too, including the way the clouds behaved differently above the warm air rising over land than they did above the cooler water.

Together, these signs made the tiny speck of an island appear to expand for the navigator, becoming detectable even though still well beyond visible range, allowing a navigator to find it in the vast open Pacific. We are interested in water, not land, but the same principles can apply. We can learn to spot the signs that indicate that water cannot be far away. This is a rewarding habit to develop, one that makes it possible to pick up the "ripples" water sends out through the land. Once you learn to sense these signs, then the small, beautiful lake off the beaten path, the one that most pass by obliviously, will radiate signals, sending out its ripples and drawing you closer to investigate.

EVERY WILDFLOWER, tree, and animal will be found with varying probabilities according to the amount of water in the local area. In the case of insects these ranges can be minute. Many insects will not be found more than a few yards from freshwater and we will meet some of them later. Common flies are rarely seen as better than a nuisance, but on a hot summer's day try to notice how their numbers fluctuate as you travel past water sources. In the Sahara I found them to be a treasured clue and one of the most dependable signs that an oasis was nearby. Bees can also be helpful, as they will often fly hundreds of yards in a straight line to and from water, forming a faint marker in the sky that points the way to it.

Birds have no sweat glands, so they lose water at a slower rate than many mammals, which means they can travel much farther than some insects and mammals from water, but never deliberately beyond certain ranges. Large birds, birds of prey, and those that feed on carcasses, like crows, get a lot of water from their food and so don't need water nearly as regularly as birds that feed on seeds, like pigeons, doves, chickens, swallows, and swifts, which will visit water regularly when they are feeding. As well as learning the way each bird relates to water, it is possible to pick up clues from their behavior, too. If you see birds flying fast and low they are more likely heading toward water, but if they are flying from perch to perch among the trees, this is a possible sign that they are full of water and are flying nearer the higher end of their weight range.

Many birds have quite specific habitats: Kingfishers are territorial river birds and return to their home patch by the river, so they are a guarantee of a freshwater river nearby. Bank swallors are another clue that should stir suspicions of a river nearby. Many birds have very distinct preferences regarding freshwater or salt water. Puffins, for example, have no interest at all in freshwater, and coots are equally uninterested in the salty stuff.

Trees, like most other plants, are rooted to the spot, which has serious consequences for them and is equally enlightening for us. Tree roots must achieve a tricky balance from the ground they will forever lie in: They must support the tree, perhaps against very strong forces like gales, they must provide minerals, and they must supply thousands of tons of water. It is this precarious balance between needing a firm anchor and needing copious water that means trees find their niche by specializing. Beech trees have evolved to tolerate much lower levels of water in the ground than most other temperate trees and indeed have roots that will not tolerate long periods of being immersed in water, which gives them a huge advantage in places where water is in relatively short supply—beeches are therefore a good sign you are on dry ground. Willows and alders will only grow well in places where their roots will be regularly wet, and so they are a strong indicator that water is not far away.

Every single one of the lower plants, from the uglier weeds to the most beguiling wildflowers, will have its preferred level of moisture and so reveal the water in the ground and therefore the likelihood of it nearby. Many aquaphiles betray their preferences in their names, from the cheery yellow marsh marigold to the nodding heads of water avens.

*Willows lining the banks of a river.*

When it comes to finding, mapping, or predicting where the water is in your area using plants, the trick is not to launch yourself into a masochistic ritual of learning the names of those plants that indicate water—there are hundreds of them—but instead start to take an interest in how the plants change as you approach water in an area you know well. You will then begin to build a collection of water-loving friends, recognizable on sight, and the names will follow this familiarity in good time. I rail against the naturalists who think that knowing the name of a plant is better than knowing its character.

WE EACH HAVE a basic, unconscious ability in this area, one that a modern lifestyle may undermine, but one that is fundamental enough to survive a tsunami of daily emails and screens. From a young age we see the grass of lawns browning in long summer dry spells and then returning with a vibrant verdancy when the rains return in earnest. It is not such a great leap from this observation to noticing that the brown summer grass grows greener as we close in on rivers.

A few months ago my elder son volunteered, hesitantly, to accompany me on a trip to Winnall Moors in Hampshire, on England's southern coast. Our joint mission was to find some signs of otters, and we did spend a fair deal of time sniffing suspect lumps in the hope of the jasmine tea scent of otter spraint. We did not have much luck on that occasion, but during an idle moment between two poo investigations, I pointed to the end of a long lane and asked him what he thought was at the end. I was pointing down a long, straight path that we had not been down, and all we could see at the far end was a tall wall of brown reeds. Reeds are a certain sign of water, but he had no way of knowing this that I could think of.

"The river?" he said, which I was delighted by, taking it either as evidence that these are skills that we pick up subconsciously fairly easily, or that the poor child was being subjected to more of this way of looking at the world than is fair for a lad of the Xbox generation.

Without parental coercion this is still a habit worth nurturing for ourselves. It will yield countless small joys, times when a visit to a local river or lake will be accompanied by noticing the change in plants as you approach and then finding yourself suddenly drawn to befriending one new example in particular. I will never forget the pleasure it gave me in learning to associate a strange late-summer snow with the nearby sluggish streams. The black poplar is the most endangered native timber tree in Britain, according to the UK's Forestry Commission. The first time I encountered this rarity was thanks to its extraordinary seeds. A carpet of white cotton wool was spread across the dark mud of a country lane, a lane that ran next to a slow stream. The next two times I came across these fluffy white seeds I noticed the same thing—water nearby—and so it was only a matter of time before a fondness for these windblown snow-cotton seeds and a curiosity about their origin drove me to associate both the seeds and their parent tree with water nearby.

Early on you will likely want to focus your efforts on the biggest, boldest signs, the ones that can be read from farthest away and with greatest ease, a line of willows indicating a river in the middle distance, perhaps. But with time it is likely that you, like me, will take more and more satisfaction in the subtler signs. Lichens are sensitive to many things, including moisture, and there is one in particular in the UK that is a strong sign that water is nearby. Its name among those that don't get out enough is *Fuscidea lightfootii*, but the rest of us can refer to it as the "lightfoot lichen." It is fairly easy to recognize, being a far-from-shy shade of bright green, with a scattering of black splotches on it. Lightfoot likes damp air and moist conditions and thrives near water. In the US, rim lichen tends to grow on driftwood and is therefore a clue to water nearby.

At the finer end of the animal scale we can take an interest in the habitats of the insects we see. Recognizing flying insects on the wing is challenging, but killing them just so you can get to know them is a bit brutal, and netting them is a bit, well, overeager. Nature solves this problem for us; just take a peek at the insects you spot in spiders' webs. They will be very different a few yards from water compared with a mile from water, and there will be gradual changes in between.

We rely on our eyes for most of the signs that we are getting closer to water, but it is so satisfying when the other senses can help that it makes it well worth cultivating them. The smell of the sea is the best known of these experiences, but only because it is the sledgehammer. More satisfying are the much fainter scents of streams on a light breeze or even those times you walk from dry ground into an area that has recently experienced a very local rain shower. Rain churns up plant oils into the atmosphere and activates the actinomycetes bacteria in the soil, and this is part of that unique smell of rain on dry ground we come to know so well. If rain falls after a long dry spell, it generates a particularly strong scent known as "petrichor."

We will delve deeper into some of these methods in later chapters. In the meantime, make sure you listen for the popping of Himalayan balsam along river banks, as its seed pods burst in the final quarter of the year. The pop is usually triggered by touch, and the small explosion will be felt as a mild sting if you are the one to trip it off; this force will fire seeds over seven yards away and into the water itself, which is one reason it does so well along riverbanks. Himalayan balsam is a non-native plant that is loathed by some, as it is such a rampantly successful invader. As a water reader, there is no need to be drawn into its virtues or perils. We can enjoy its purple flowers and the popping sound that indicates the likelihood of water close by.

In hot, dry parts of the world, explorers have long learned the plants that indicate water (which is most of them in the driest deserts), but also the plants that hold water. The traveler's palm aligns east–west, but may have earned its name more as a reflection of its habit of holding water in the base of its palm fronds. Closer to home, there is some fun to be had in knowing the plants that will do this, too, even if it is not necessary for survival. Teasel is a common plant that you will have seen many times, even if you don't know it by name. It frequently grows to six feet or more and is recognized most easily by its prickly stem and leaves and especially its head, which sports pink/purple flowers in summer, but then leaves a trademark brown head for the rest of the year. Teasel is one of those plants where the Latin name is enlightening. Its genus name is *Dipsacus*, which means "thirst for water," and this refers to the way water collects in tiny cups at the base of the leaves, where they join the main stem.

IF YOU ARE struggling to locate water in a city, but are craving the sight of some, then a little lateral thinking can go a long way. Single-engine helicopters tend to head calamitously quickly toward the ground if they suffer an engine failure. This means that it is conventional (and often legally enforced) for these helicopters to stick to routes that minimize the risk of coming down in densely populated areas in the event of an accident. But land in cities is by definition densely populated with only a few exceptions, and one of these is any river that passes through the city. Watch the helicopters over a city like London and you'll notice that many of them follow a sinuous curve through the middle of the city. These helicopters are painting the line of the Thames in the sky for you.

Water has a habit of forcing its way into place names, so streets with names that are obvious, like Bridge Lane, will be joined by other place names with slightly less-blatant clues within their titles: bourne, burn, brook, strath, mill, gill and many more, indicate the likelihood of water nearby. Both "aber" and "inver" stem from Celtic words meaning the mouth of a river, or where waters flow together, so it is no surprise to find that Abergavenny and Inverness have their rivers.

Together all these techniques can be honed into the fun art of learning to see the water, before you see the water. My final tip would be to retrace your steps occasionally. If you find yourself coming across water when you weren't expecting it, then this is a great opportunity to go back and walk slowly toward the water again, only this time super-tuning your senses for the clues in nature that water is nearby. Cheating by knowing that the water is definitely there is one of the best ways of honing your skills to the point where the water will find it hard to catch you off guard in the future.

# 4

# The Not-So-Humble Puddle

THE PUDDLE'S DOWNFALL lies in its humility. There it sits, low down and apparently motionless, meekly refusing to seek our attention. The only time a puddle makes it into a story is when a car goes through one, and then it's all about the rude driver. The puddle itself gets ignored. No longer!

A puddle is evidence: There is quite a lot of water in one particular spot, but none all around it. Why? Puddles are not random. In this chapter we will look at different types of puddles, which all form part of a family tree consisting of the Low Point puddles, the Tracker and the Navigator puddles, the Overhang, Spring, and Seismometer puddles.

Every puddle is a sign that the water has been blocked, stopped from traveling down through the ground. So if a puddle is persistent, then the first thing we can deduce is that the ground beneath the puddle is either not porous or it is saturated. This is mainly interesting when we travel through a rural area and notice that the number of puddles suddenly increases, despite there not being any more rain in that area. This is a sign that the rocks beneath your feet have probably changed, even if the appearance of the mud has not changed. Since the rocks are responsible for a lot of the characteristics of the soil in an area, and the soil strongly influences the types of plants and animals you will find, a sudden change in the number of puddles, without a very local downpour, is a sign that the rocks, soil, plants, and animals all around you will also have changed.

When focusing on a particular area to ask why there are puddles in some specific places but not others, we quickly realize that every puddle is indicating a low point in the local landscape. Water is being pulled down by gravity, so it is always trying to travel downhill until something stops it. For this reason, any local low points of non-porous ground will host puddles.

Roads are designed with a camber so that the water flows from the center to the edge of the road, precisely to avoid puddles in the middle of the road. This water then collects at the edge of the road and should continue to flow gently dowill into a drain. But time has a tendency to warp and buckle the neat plans of road builders, and very often we find that this smooth line from the center of the road, all the way to the drain, gets bumped, bent, and dented by cars, people, and ice, to name a few. Sometimes the place that semitrucks like to stop and unload leads to a depression and a puddle.

Whenever roads are dug up, for repairs or to lay cables, the road is sealed over, but nearly always with a material that is different from the one originally used to build the road. Over time this will swell and contract at a rate different from the rest of the road, and it is very common to find a puddle at the join of the old tarmac to the new. Sometimes the drains themselves cease to work, an urban equivalent of a non-porous rock that stops the water continuing on its downward trajectory, and that will create a puddle, often a very large one.

Urban planners and road builders start with an aim of eliminating puddles altogether, so a puddle anywhere in a town is a sign that something has gone a little bit wrong somewhere, and the simple story behind that can usually be found. From obvious observations, small insights can blossom.

Low Point puddles are very common, but not all of them are fascinating. We find a much more interesting branch of the puddle family if we appreciate that the reason one tiny patch of land is lower than the ground around it is because it has been eroded by something. At this point the puddle becomes a clue to activity of some kind, and this means it is part of the Tracker family of puddles. Tracker puddles are a collection that will give you some clue as to those who have been there before you, and what they have been up to.

Whenever anything travels over the land, from a bicycle to a beaver, it will leave tracks, and if these are more common in one place than another, then the ground becomes worn and a small depression may be formed by this erosion. After rain, this depression will collect water. At the most dramatic and obvious end of the scale, water will collect in the ruts and ridges caused by tractors and form a puddle—we've all seen thousands of these puddles and stepped to avoid them. But not all puddles have such an obvious history.

Where two paths or tracks cross there will be a patch of ground where the land is worn much more seriously than on either of the two paths, because all the traffic following each path must also cross the junction. This leads to double the erosion in that junction "box," and so the ground gets worn down and a depression, or more likely a series of depressions, forms at the junction. So, whenever you are on a path that reaches a junction, it's worth a short pause to look for the Junction puddles, part of the Tracker family, and usually nice and easy to spot.

Now, notice how turning creates more wear and tear on the ground. Turning is something that uses a lot of force, and this force wears away at the ground. At every junction, there will be evidence of the way things have turned. Often there will be curved puddles at junctions: big, glaring ones if vehicles have been turning, slightly subtler ones if cyclists have been turning, and subtler still if walkers have turned, but evidence of each is usually there to be found. Working out the most common direction that people turn is usually fairly easy from the shapes in the mud and the puddles. Whenever a minor path meets a major one, I make a habit of trying to work out the most common direction people turn, as this is the probable direction of a town or village. A Turn puddle is a specific and curved type of Junction puddle, and both are part of the Tracker family.

The concept behind Tracker puddles is simple: The more erosion from somebody or something traveling, the more likely there is to be a puddle in one spot. But just because it is simple does not mean that it cannot be beautiful, and the lighter the footprints, the more joyous it is to spot the puddle they create.

The next time you are walking along a country path and you find a puddle on an otherwise level stretch of mud, pause to see if you can solve the puzzle as to why it is there. Peer at the undergrowth on either side of the path and look for evidence that someone's been busy. Animals have their own network of paths, and the badger's big, straight ones are quite easy to spot, especially if you get down low and view from badger level. There are many others, though, like deer and rabbits, that will forge their own highways, and where these intersect with our own paths, it follows logically that there will be more wear to the path. Only a little bit compared to the junction of vehicles or even walkers, admittedly, but that is all it takes for one small patch of ground to get worn down a tiny bit, which in turn makes it the logical place for the water to collect after the next downpour.

*Turn and Junction puddles.*

The next thing that happens is that these baby puddles grow, helped along by two factors. Firstly, the small hollow created by the intersection of humans and animals stays soft and wet longer after rain than the drier mud around it. This means that when the next feet land there, whether the cumbersome boots of a walker or the feathery feet of a vole, they churn the mud up a little bit more than the dry hard ground to the side, and this erodes the land more quickly in that one small spot. A self-reinforcing cycle has begun, and the puddle grows a small amount.

Secondly, all small puddles act as miniscule reservoirs for the animals in that habitat. Thirsty animals will seek out the puddles, just as they travel to the larger ponds and lakes, and this leads to yet more footfall, gently encouraging the infant puddle to grow. My dog regularly weaves out of his way to slurp at the puddles on our walks, especially later on in the walk when his thirst grows, and he is just one of millions of animals using these natural water fonts.

On first reading this, you will doubtless be a bit skeptical that the dainty prints of a rabbit can actually lead to a puddle. But they do, and the key is that we tend to think of the single instance of the animal crossing paths with us, but patterns in nature are usually created by cycles and repetition. One rabbit crossing our path may not create a puddle, but a few doing it several times every day for months may well. And for the reasons above, once the tiniest puddle is born, it is more likely to grow than fade away.

These Tracker puddles are everywhere, once you learn to look for them. They are one of the reasons that a puddle should not be viewed as a random mild hazard, but instead a small clue as to what is going on around us. If we want to add a few layers to our understanding of nature, then the act of pausing to consider why the puddle is there will often open up the next level, because around the puddle we may notice the footprints of the puddle's creators.

The soft mud around most puddles is an ideal spot to look for the animals' tracks, at which point the story can become much more detailed. This is not a book about tracking, but it is worth our dipping into a little basic tracking each time we see a muddy puddle, because it is part of that puddle's story. Try to work out which animals have passed close by—you are very likely to spot and be able to identify the pads and claw marks of dogs, for example.

Next, look to see if you can work out if the puddle was the destination for that animal or not. Do the tracks pass alongside the puddle or do they point in toward it? It is very straightforward from there to work out whether the animal was having a drink or just passing by, trying not to get its paws wet. And now spend a moment looking to see if you can decipher where the animal came from and headed to afterward. Did it follow the path you are on—very likely if it is a dog—or cut through undergrowth to the side somewhere nearby—more likely if it is a wild animal?

On a walk last year, I turned a corner to find a large puddle reverberating with ripples. It was a large, typical Turn puddle, created at a dark mud junction by a farmer's tractor, and it would not normally have held my attention for long, but the ripples caught my eye. There was a calm center to the puddle, surrounded by a short series of ripples marching toward the puddle's edge. It was a very calm day, so I knew that I could discount the wind, and the ripple pattern was wrong anyway. Besides, I had a more likely suspect in mind. I dropped back, quietly concealing myself in the undergrowth in the direction I had come from and stayed perfectly still as I studied the puddle and listened. Sure enough, a minute later the culprit returned, and I spent a wonderful couple of minutes watching a nuthatch take his bath.

The ripples in a puddle will reveal things, just as the ripples in a pond and waves in the ocean will. Drop a stone in a puddle and you will be able to see the ripples charge away from the disturbance. If the puddle is big enough, then you will notice the center return to calm, before the ripples rebound from the edges and the reflected ripples return, sometimes creating diamond patterns and crests as these tiny waves crash back into each other. The brief calm in the center is the clue that whatever was disturbing the puddle is doing it no longer, the prettiest case being a bird or insect taking off.

A LITTLE OVER a hundred years ago, A. M. Worthington, headmaster and professor of physics at the Royal Naval Engineering College in Devonport, in South West England, took advantage of the new technology of high-speed photography to investigate what a splash is, how it forms, and what it actually looks like. He wrote up his results in a book published in 1908, called, appropriately, *A Study of Splashes*. There are lots of interesting insights in the book, but most are hard to see without fast photography, so I will stick to mentioning those that have some relevance to our own studies.

Worthington discovered that bubbles sometimes form from a splash, and this is related to the height that the liquid drops from. If a drop falls from high enough, a bubble may form, but below a certain height they do not form at all. He also noticed that milk and water form a different pattern of ripples when a drop lands in them and recognized that the consistency and surface tension of a liquid will give it its own signature ripple pattern. We need not study splashes in the same level of detail that Professor Worthington managed, but it may be worth spending a moment with his words, for no other reason than it will help remind us that there is an awful lot going on out there that we either fail or choose not to notice:

> Let the reader when he next receives a cup of tea or coffee to which no milk has yet been added, make the simple experiment of dropping into it from a spoon, at the height of fifteen or sixteen inches above the surface, a single drop of milk. He will have no difficulty in recognizing that the column which emerges carries the white milk-drop at the top only slightly stained by the liquid into which it has fallen.
>
> In the same way naked-eye observation reveals the crater thrown up by the entry of a big rain-drop into a pool of water. In either case what we are able to glimpse is a "stationary" stage. The rebounding column reaches a maximum height, remains poised for an instant, and then descends. The same is true of the crater. It is the relatively long duration of the moment of poise that produces on the eye a clear impression where all else is blurred by rapid change.
>
> But there is frequently a curious illusion. We often seem to see the crater with the column standing erect in the middle of it. We know now that in reality the crater

has vanished before the column appears. But the image of the crater has not time to fade before that of the column is superposed on it.

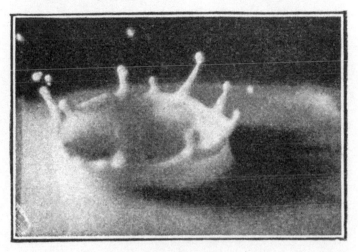

*A Worthington splash.*

What if you see a puddle that isn't obviously the result of animals or people eroding the ground and that isn't in an obvious Low Point? Check for the Overhang puddle. Rain is often collected and funnelled by trees and anything else that hangs over the ground. A good sign that you are looking at an Overhang puddle is a strip of moss indicating the flow of water from a tree or part of a building down to that spot in the ground, as moss is a guarantee that an area is regularly wet.

Some of the most interesting puddles have as much to do with the sky as the ground, and to understand these we need to consider the way the water arrives as rain and is then dried by the sun. This leads us to the Navigator family of puddles. There are two casual and very widespread assumptions that the successful puddle reader must banish. Contrary to popular imagination, rain rarely falls perfectly vertically and, outside of the tropics, the sun is never overhead. The heaviest rainfalls are usually accompanied by strong winds, which means that water will be driven by the wind and hit buildings, trees, and hillsides at an angle. This will lead to water collecting more voluminously in some places than in others. If you have been tuned to the wind's direction during the rainstorm, this will make interpreting these puddles and using them to deduce direction quite straightforward. Over longer periods the rain will arrive mostly with the prevailing winds, which means that it collects and pools on that side of any obstacle, which in turn leads to a greater prevalence of puddles on that side. However, these puddles tend to be short-lived, because they are usually dried quickly by any afternoon sun. The cousin of the wind-driven rain puddle is the snow or ice-melt puddle. Snow drifts and accumulates in certain spots, and when the temperature rises, cold puddles remain.

In my earlier books I have gone into some detail about the direction of the sun at different times of the day. It is a critical part of a natural navigator's awareness, but when it comes to puddles we can keep it simple. For everywhere north of the tropics, which includes all of Europe and the US, the sun is due south in the middle of the day. The sun is giving us most of its heat and light at this time, and so this is when it does most of its drying. This means that anything shaded from the southern midday sun will take longer to dry, and over time this often leads to more puddles on one side.

That is hopefully simple, but it does lead to some slightly surprising results when it comes to using puddles for navigation. When an obstacle is tall, like a building, things are straightforward. You get more puddles and longer-lasting ones on the shaded north side. The next time sun follows rain, look at the roads of any town and you will see this effect in action. The tarmac and pavements exposed to the sun dry much faster than the shaded north side of buildings; on a cool day you will sometimes see the steam rising from one side of the road only. This leads to more puddles on the north side of buildings.

*Puddles are more common on the south side of tracks.*

When obstacles are a little bit shorter, like the undergrowth at the side of country tracks, the same rules apply, but this leads to a result that is a little counterintuitive for many rural natural navigators. The puddles are still on the north side of the obstacle, but this means they end up on the southern side of the track. Most people might expect to see more puddles on the northern side of a track, but the southern side is the shaded side in this case, as the illustration above demonstrates.

Using puddles to navigate is a lot easier than many would imagine. It just requires an understanding of where the wind and rain have been coming from and remembering that the southern sun will dry the ground asymmetrically, leaving puddles to the north side of any shadow-casting obstacles.

Around the edges of puddles you will find that life changes a little, just as it does around all water. Grasses and weeds that cannot survive a few feet away may thrive at the edges of long-surviving puddles. You will find many insects on or near puddles and occasionally come across life in the puddle itself. In dry areas, puddles will sometimes become home to frog-spawn, but sadly not many tadpoles make it to become frogs in these puddles. The tadpoles that hatch will eat any available algae, then, starving, they turn cannibal and eat each other.

The biggest life form you are likely to find near a puddle is the human being. A few puddles, sometimes very large ones, are created by water from below rather than above the ground, and these Spring puddles were once a key source of fresh water. Where a porous rock, such as chalk, meets a layer of impervious rock, there will often be a spring, and where there is one there are likely to be lots more at the same height.

If it is an open dry landscape, these springs are usually easy to spot, as they make the land around them more lush and verdant than the surrounding countryside. Look at a good map and you will occasionally see a label for a spring. If there are few other sources of fresh water nearby, it is very likely that there will be a village or other civilization near these Spring puddles. Humans congregate around precious fresh water, just like flies.

In wet parts of the world, like Britain, this strong relationship between human habitation and bubbling springs is often overlooked, but in slightly drier parts of the world, like southern Europe, the evidence is vibrant and stark. The writer Adam Nicolson found that in Greek the word for springs has become intertwined with the meaning of vegetable gardens. In Greece, to say you are going to the *vryses*, the springs, means you are heading not just to the water but to the place of growth, food, and life.

There is something pure, primal, and exquisite about drinking spring water. Even gin-clear streams may deceive, and more than once I have rounded a bend of temptingly transparent water to find a rotting dead sheep or other putrefaction in the water. But a spring will not and cannot trick in that way. The water falls as rain, gets filtered over many weeks by rocks, and emerges at your wild table not just gin clear, but virgin clear.

"ARE YOU OK?" the short traffic cop asked with parking ticket machine in hand, which was more kindness than this species had shown to me in my life up to that point.

"Yes, thanks. Very well," I replied and returned to my work. I pulled out a camera, not that I needed or wanted one. I have just learned that it calms people down in these situations and saves a long explanation that is not going to be believed anyway. My belly was now cold, after several minutes lying on the pavement near the Harrods department store in affluent Knightsbridge. Perhaps Knightsbridge was not the best location for this experiment. I probably looked like the world's laziest terrorist, but then I wasn't sure anywhere would be perfect. After much wriggling and squinting I stood up and stepped back, then squatted and peered. Finally, I found what I was looking for.

The Seismometer puddle is the puddle we can use to measure the slightest movements in the ground or air. Navajo Indians were reputed to be able to discern whether horses were approaching, even the number of horses, their speed, distance, and whether they had riders on, just by placing their ears to the ground. The principle behind a Seismometer puddle is very similar. By tuning to fine vibrations in the ground, we can learn to predict things that the urban natives around us fail to spot. Like whether our bus is on its way or if the train has just passed under us. To understand how this puddle works, we need to think about binoculars for a moment.

Have you noticed how difficult it is to keep a pair of binoculars fixed on a very distant, moving small object, like a bird, and how every time you breathe you seem to lose sight of it? You are probably aware that competitive rifle marksmen give as much thought to their breathing as to their rifle. The reason for both of these, is that the further something is away, the greater the effect a small angle change will have on what you see.

Coming back to the puddle on the pavement, if we look at our own reflection, it would take quite a major disturbance, a foot stomping in it or a strong gust of wind, to have much impact on that image in the puddle. But if we look at something very distant in the reflection, we will pick up much more subtle angle changes, which means that much tinier movements in the water's surface become apparent. And the very best situation is when we line up something small, bright, and far away.

One late spring dusk, walking along a country lane, I noticed that both Venus and Jupiter were visible in the puddles I passed. Seizing this Seismometer puddle opportunity, I made myself comfortable by leaning against a tree and began staring at the image of Jupiter in one of the larger puddles. Nothing happened for a few minutes, but then I noticed a faint disturbance in the water. A few seconds later it happened again.

At first I thought it must have been a small insect, but I'm familiar with most of the patterns they create and it didn't quite fit. The pattern disappeared for a couple of minutes, but then returned again. This continued for a while, and then finally I spotted the source of the tiny disturbance. A bat was swooping over the water, and the faintest breeze from its wings was visible in the motion of Jupiter oscillating in the puddle.

Puddle reflections are underestimated, but there are at least a few who have sought their potential. The photographer Brian Podolsky captures the world seen through the reflections in puddles, an art he calls "puddleography." He claims that puddles offer "a window to another dimension." I'm not sure about another dimension, but if you seek out your own Seismometer puddle, find a distant object in the reflection. Then you will be ready to detect the flight of a bat, an invisible train, or four distant galloping cowboys.

# 5

# Rivers and Streams

THERE WAS AN ATTEMPT in the 1920s to classify river stages according to the fish that lived there, but it was only partially successful. From the fishless highest mountain streams, down through brown trout becks and on to minnows and then narrower fish like sunfish and bluegill. It works in a loose sense, as fish have their habitats, but unlike plants the fish refuse to stay put! Depending on whether the expert you talk to is a hydrologist, geologist, angler, or entomologist, they will choose to define river stages in their own different ways and from a very long list, which can quickly become confusing and unhelpful. Even mathematicians have their ways of labeling river stages and behaviors and have developed a formula, called Manning's Formula, that supposedly takes everything into account and describes the velocity of the river, clarifying very little for us in the process. Fortunately we are going to simplify things by deciding that a river is either in its upland stage or its lowland stage.

Generally speaking, an upland river is steeper, as the higher we go the steeper the land gets on average, so the water is more energetic. Some lowland rivers move at such a sedate pace that you can walk alongside one, following its flow for half a mile and the river will have dropped by less than half a yard in that time. This makes the diagnosis straightforward: if the water moves rapidly at times and carves narrow, sometimes steep channels, then it is an upland river. If it is broader, slow, and forms wide meanders, then it is a lowland river. Depending on your diagnosis, there are some specific things to look for. An upland river carries everything from large boulders to gravel along with it. Take a look at the sides of the river and across its flow and gauge the rough size of boulders that have been moved. This gives you a vague estimate of the power of the river when it is full and roaring. This is the second thing to sense—can you hear the water? Upland rivers have lots of white, breaking water, even when they do not appear especially steep. We're not talking about waterfalls here, just water moving fast over rough, hard ground, so that it gets churned up, turns white in places, and makes a familiar sound. It is the air mixing with water that actually creates the sound of moving water. Lowland rivers are much quieter, the water itself often near silent.

If you look closely at some of the flatter boulders in an upland river, you will be able to spot two more clues related to the water's action. Notice how you can often find little pools forming in the tops of these broad rocks and how others have a pock-marked surface. These are two signs of turbulent erosion, as gravel gets whipped around in a vortex when these rivers are in full flow and carves these holes and dents in the rocks. These then fill with river- or rainwater, creating mini-pools in the rocks.

Next, have a look just downstream of the larger boulders in the main stream and see if you can find some "lee scree."

Whenever a fluid carries particles past an obstacle, we get clues to the flow from the patterns it leaves behind. In natural navigation, it is very common to search for patterns left behind when the wind deposits snow, sand, dust, or leaves on the sheltered side of any obstacles in the path of the wind. This is helpful because once you know which direction the wind came from, these little tails of particles form an easy compass.

The same thing is happening in a fast-flowing river, only here we are using it to deduce the strength of the water, rather than direction. All flowing water gathers particles of varying sizes and when this water gets slowed by any obstacles in its path, like a boulder, the particles will get deposited on the lee, down-flow side of the rock. The size of these particles, from sediment up to rocks, gives a clue as to how fast that water must flow at times, because a weak trickle can carry silt, but only a tumbling cascade of water will shift the bigger rocks.

The shapes of the rocks you see in a river and by its side are also a testament to the action of the water. Rounded pebbles must have been carried by water, and the smoothness and evenness of this rounding is a clue as to how much they have been eroded by that water. Rocks that are sharp or angular have not spent time in moving water. Think of the sharp, broken glass of a smashed bottle left on the beach. If it stays on the high, dry part of the beach it will stay sharp for decades, but if it ends up in the sea, it will be transformed into the smooth, rounded glass pebbles we sometimes find in the sand.

If you are looking at a lowland, meandering river, there is a fair chance that the water is not clear enough to peer into— because it has gathered too much silt and become opaque. So the clues to look for in these rivers are on the surface. We will be going into lots of examples of those shortly, but for now just notice how the water moves more quickly around the outside bend than it does around the inside one.

## FLASHINESS

If you have come to know any river well, then you will have learned to gauge how your particular river responds to fluctuations in the weather. But, whether we know a river well or not, there is a risk of making some general assumptions that are a bit simplistic. If we do know a river, we tend to assume that all rivers will behave like that one, and if we don't know one we lean toward some seemingly common-sense assumptions that can prove surprisingly wrong.

For example, what will happen to a river after lots of rain? Surely it's straightforward—it's just the basic hydrological cycle: Water evaporates under the sun's heat, condenses into clouds, falls as rain, and finds its way back to the rivers—so, the more rain, the higher the rivers! That is true up to a point, but the curveball in this simple equation is that the time that it takes a river to react to rain varies wildly depending on the land around it. After a period of very heavy rain, some rivers will sweep cars and even trains away, whereas others will barely register a rise. Why is that?

The answer comes in the word "flashiness," which is the term used to describe how dramatically a river responds to rainfall. This is not a casual term; it is taken very seriously by hydrographers, because it is hugely helpful in both measuring and then predicting the way a river will behave.

Let's look at a couple of actual examples. If a river is surrounded by impervious rocks and soils, like clay, then pretty much every drop of rain that lands and doesn't evaporate has nowhere to go but downhill until it finds a stream and then a river (a stream is just a river you can step over). But if the same amount of rain falls on a porous rock, like chalk or limestone, then it will sink into the ground and keep on going until it hits something impervious lower down. At that point it starts building as part of the "water table," forming an underwater reservoir known as an "aquifer" in the porous rock.

This water doesn't see the light of day again until it emerges as a spring, often a long way from where the rain landed and typically much later, often months after the rain fell. This is why anglers who favor chalk streams in summer have a saying, "The only useful rain is that which falls before St. Valentine's Day." Anything that falls after February may only reach the river after the fishing season ends in the autumn. So rain that falls on clay will send the local river levels up in hours, whereas the exact same rainfall on chalk will have barely any noticeable effect on a local river for months.

How can we tell how the river we are looking at will behave? Will it rage after rain or barely register a change? There is a nice, simple clue in the shape of the bridges you see over rivers. Flashy river levels rise so quickly that any bridge that didn't take this into account would be swept away in its first winter. Bridges in flashy river country are higher and have tall supporting pillars, while those in un-flashy areas are lower with modest pillars. So a quick look at the height difference between the top of the water and the bottom of the bridge is, other things being equal, a guide to what that river will do after heavy rains. One of my local bridges, Houghton Bridge, is very low, because the river it goes over, the Arun, is in chalk country. This surprises some people because it is in a part of the world that floods almost every winter. It may flood, but the water speeds and levels are very slow to rise, so the bridge can afford to be low.

Floods can happen almost anywhere; it is the speed at which the water rises that is often more important in terms of the effect that water will have. Some of the most dangerous floods happen in the places they are least expected. A "wadi" is usually thought of as a dust-dry ravine in desert regions. Wadis are sought out by desert travelers for the plants that grow in their lowest parts, offering precious food for camels and other animals. However, the reason it is a ravine at all and carved lower down, nearer the water table, than the surrounding land is because this is where the flash floods will surge when the infrequent rains do come. Isabelle Eberhardt, the extraordinary cross-dressing Swiss explorer, once said:

A nomad I will remain for life, in love with distant and uncharted places.

But a place can be uncharted and still reveal its secrets. Desert wadis are places to be wary of after heavy rains. Eberhardt died in the Aïn Sefra wadi in Algeria during a flash flood, at the age of twenty-seven.

Cities can appear impervious to water, which doesn't sink down through the roads or buildings at all, but still needs somewhere to go. The drains and sewers in London, many dating back to the mid-nineteenth century, struggle to cope with the heavier rains, and so even if there are not flash floods, there are pungent odors near the Thames at times, a form of nasal flashiness. In towns it is always worth looking for plaques that mark the high-water level of floods in years gone by, as it is a custom to mark the historical high points in places that flood, a practice that stretches back to ancient Egypt. These marks can help us to build a picture of the river's habits, but don't fall into the all-too-common historical trap of thinking that these mark the limits of what is possible. Rivers like to break records, and the flashier they are, the more violently they will do it.

In the countryside, plants make their own flood markers. An area around a river devoid of scrub, with just grasses thriving, is a sign of earlier flooding or active grazing, or both. But some plants can be quite specific: Reed canary grass prefers areas that alternate between wet and dry.

## THE WATER TABLE

In my experience, dowsers are passionate and honest people, but there is a common, fundamental misunderstanding about dowsing and underground water, and it is very important in the context of river levels.

As I mentioned earlier, water that falls as rain will gravitate down through the ground until it is stopped by something, typically impervious rocks, where it will collect. This forms an underground reservoir, known as the water table.

The water table refers to the level of saturated ground. It can be visible in the case of rivers, but usually refers to the invisible reservoir of waterlogged porous rocks in the land. The level of the water table rises and falls with rainfall. The water table will be found above underground impervious rocks almost everywhere, even in deserts. This means that you are likely to find water almost anywhere you choose to drill down. The only pertinent question is usually: How deep is it? So when dowsers successfully find water with divining rods or any other device, this is only impressive if they have specified the *depth* of the water. Blindfold yourself and throw a dart at a map, then start digging where it lands, and you will likely find water *if you go deep enough.*

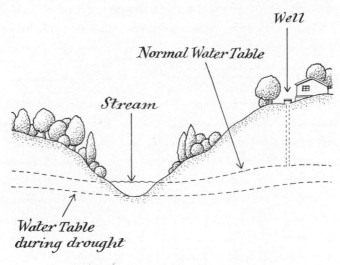

*Predicting the depth of water using nearby rivers.*

How can we work out how deep you would need to go? Rivers can give a good clue to this, and it is very helpful for any well-diggers or wannabe dowsers. All rivers have levels that fluctuate, but many have a base level that they rarely drop much below. This is the level that you will typically see in middle-to-late summer or after other long dry spells. It is a clue to the height of the summer water table around you. (If a river dries altogether, then that is just a sign that the water table has dropped below the height of the riverbed. Conversely, if the water table rises until it meets the ground all around you, the result will be wetlands.)

So if you want to predict a water "strike" better than a dowser, just find the local river in summer and look at how much higher is the ground you're interested in than the level of this river. This is roughly how deep you will need to dig before things get very wet.

## PLANTS AND ANIMALS

The following observations will help you to associate the plants and animals you see with the conditions in the water around you, which can culminate in some quite specific deductions. If the water is clear, find a bridge and have a look at the life you see either side of it and compare this with what is growing under the bridge, both in the water and either side. This will underline something we know to be true on land but often overlook when peering in water: Plants need light. You will find different and fewer plants in areas of shade—even a thin tree cover can be enough—compared with plants even a few feet away, in the sun. In light areas you may spot common water "weeds" like white water crowfoot, also known as water buttercup (as it is part of the *Ranunculus*, buttercup, family) easily recognizable in summer from its carpet of white flowers with yellow centers. But whatever species you see, note how it is either a lover of sun or shade. Next note how fast the water is flowing past these plants. White water crowfoot, native to the western US, will tolerate water up to a certain speed, but then gets washed away, so don't expect to find it in upland streams.

If you see snails clinging to the surface film of the water, then it is very likely that the water has warmed up a fraction and the oxygen levels have dropped to a level that is starting to threaten life in that water. The poor snails are gasping for breath and in doing so are giving you a clue to the temperature and gas levels in the water.

If you spot a film of green on the water that on closer inspection is made up of thousands of tiny floating plants, you are probably looking at duckweed. It thrives in still or very slow moving water, especially when the water has been enriched in nutrients. The feces of waterfowl like ducks enrich the water with nutrients, so the duckweed, the ducks, and the slowness of that patch of water are all connected, and "that patch of green down there" becomes part of a more enlightening natural map of the water, its quality, and its flow. There is a small delight to be found in mapping a thin stream of faster water that runs bare through an area of duckweed.

Water lilies are rooted to the bed of ponds and slow-flowing rivers and give a good insight into the conditions around them. White water lilies like shallow, very slow, and very clear water, so are more common in ponds than rivers, but if you do spot them in rivers you are looking at very pure water that is relatively undisturbed and not more than two yards deep. Yellow water lilies indicate water that is perhaps up to five yards deep and will tolerate more flow in the water. However, neither white nor yellow lilies will tolerate the sort of disturbance that boats create, so they indicate the water around them is free from regular boat traffic.

WATER MOVES MORE QUICKLY around the outside of any bend in the river. This leads to erosion on the outer bends and the deposition of sediment on the inside bends. This in turn leads to plants losing a battle on the outer bend and winning one on the inside, which means there will often be big differences in the plants you will see there. The inside bend is the newest and usually very fertile land, too, so it is often rich with the pioneer species like hemp agrimony (found in the eastern US), willowherbs, and young willow trees.

The animals that live in or by the river are not only dependent on many of the plants that grow there, but also sensitive to the speed of the water. The American dipper, a small, plump bird, gray all over but its white eyelids, will only ever be found in or near water that is fast-flowing. Native to the western US, it is well known for walking into and indeed under the fast water. The nature sound recordist Chris Watson has commented that the dipper has a song, "the pitch of which is high above the fundamental frequency of a fast-flowing stream and a wonderful example of a song that has evolved alongside running water."

The dipper is not the only animal to prefer life around fast water. There are others like the red-breasted merganser duck and the common sandpiper. Then there are animals that will only be found by the slower waters of lowland rivers, including birds like coots, mute swans, gallinules, geese, and cormorants, as well as many of the dragonflies and damselflies.

But the animal tapestry gets richer still because damselflies and dragonflies (identification tip: if the body is thicker than a matchstick, it's a dragonfly) are more likely to be found in sunlit spots than shady ones, so these insects are mapping both slow water and light levels for us. The creatures you see in the water will give you a good clue as to the minerals in that water and therefore in the land around. Crayfish need a lot of calcium to build their shells, so they are a sign of chalk or limestone in the land.

Joining two of the ideas in this chapter together demonstrates how interconnected all of these things are. The sight of a crayfish indicates the chalk in the surrounding area, which in turn tells you that the heavy rains will not lead to flash floods.

THE ANIMALS AND PLANTS we see will also be shaped by the seasons and much shorter time cycles, too. On rivers that are tidal—these are the lowest of lowland rivers, ones within reach of the sea's cycles—the life you see will flex with the ebb and flow of the water, too. Birds like cormorants prefer to fish on the ebb than the flood. There are some beautifully esoteric animal cycles out there; eel migrations are sensitive to water temperature, moon phase, and even atmospheric pressure.

Fish have been used as a guide to water quality for centuries, both salmon and brown trout being an excellent sign. But they are a relatively slow and crude indicator, as the insects will react more quickly and sensitively to many environmental changes.

It is an art that can be refined in the case of the brown trout's fins, however, as these were once used to indicate lead pollution in Welsh streams—the more black in the trout's tail, the greater the lead levels in the water. It was so effective that it was even given a name, "Black Tailing," which was used to spot where lead from disused mines was making its way into the water.

Since the fish, insects, and plants are all dependent on each other and the water itself, the simplest possible guide to the health of a river is variety in what we see.

## MOSSES AND ALGAE

There is a temptation when walking alongside upland streams to see green patches in the water as one, random, amorphous collection of green stuff, but that is to miss a water-reading trick. It is usually very easy to differentiate between mosses and algae, even though there are thousands of species in theory. This is because we are used to seeing and recognizing mosses on land, and their form in streams is not that different, clumps and carpets that cling closely to the rocks they are on. Algae can come in many different forms (including seaweed) but in streams it is often filamentous and has a recognizable flowing hair-like quality to it as it tails away from the rock it is attached to.

Once we recognize mosses and algae as different, it is very easy to spot how they are signs to different things. But first a couple of things they have in common: both mosses and algae need light to photosynthesize, so light levels will influence their growth significantly. (In the summer I can estimate roughly how long I will need to spend brushing algae in the pond at home each weekend by how much sunshine we have had during the week.)

Both moss and algae in a stream are signs of fairly continuous wetness as neither can thrive in dryness; moss cannot reproduce if it dries regularly. Sphagnum moss—spiky, spongy carpets of varying hues—is among the most sensitive, and so is a sign of permanent wetness. Notice how you only get moss and algae in the parts of a stream where the water levels are dependable; either side of the main, all-year stream you will often find the places where the water from a heavy rain flows, but then these channels will dry for long periods. Notice how bare these areas are, free of both moss and algae.

The next thing to notice is a difference. Moss will only do well on stable rocks, ones that are not being moved by the water, whereas algae can spring up temporarily in places that are more fluid. The old saying holds true here—"a rolling stone gathers no moss"—so moss shows you the parts of the stream where the rocks are settled and is therefore often a good clue as to where to place your feet if crossing. Both algae and moss can be slippery, of course, but moss tends to be less slippery, and at least the rocks with moss are guaranteed not to have shifted much recently.

(If you do cross enough rivers and streams, you will slip in one day, and if the water is fast moving it's a good idea to point your legs downstream as soon as possible after tumbling, as the thing you want to avoid is your head hitting something. Remember, as the speed of water goes up by a factor of 2, the size of object it can carry goes up by 64.)

Moss, like most plants, is sensitive to the pH levels, and species will vary with the rocks in a river. If you notice two different species growing on rocks close to each other, look more closely at each rock. It is possible one has been washed down from an area of different geology, a sign that the character of the land and water upstream may be quite different. Algae in fresh water is a sign of nutrient enrichment. A little algae is normal, even in the purest streams, but a sudden bloom is a sign of imbalance: Something with lots of phosphates or nitrates in it has entered the water upstream, fertilizer or effluent being likely suspects. Again a little specialization can solve any riddles here, as each individual alga will have trademark sensitivities to each chemical, so the answer can be deduced if the desire is there.

## RIVERBANK SIGNS

Let's look at some more of the specific signs at the water's edge. It is worth noting whether you are in a place where cattle can reach the water, as this will have a dramatic impact on the river itself. Cows trudge down to the water and gradually demolish the riverbank, leading to wider and shallower streams, muddier water, and a change in the plants. For this reason, farmers and others usually go to some lengths to stop the cows reaching the water. As well as fences, look for ditches running alongside the river, as these will contain much slower water and allow you to compare how the plants and animals change with water speed in otherwise identical environments.

Trees have almost the opposite effect of cows, shoring up the bank and resisting erosion, so you will often notice a narrowing of the river and a slight constricting of its flow where trees appear at the water's edge. (Strictly speaking, the river hasn't "narrowed" at that point, it has just resisted widening.) If you spot a line of willows, which is very likely as they are one of the few trees to thrive with waterlogged roots, look to see if you can tell if these willows get younger in one direction. Riverside willows, especially crack willows, can propagate from twigs that get washed downstream. The clue that this is happening is a steady age difference in one direction, and they will all be the same sex, male or female, and so have catkins that are all alike.

Look out for a branch or other perch overlooking the river with white splashes on it. This may well belong to a kingfisher, and the perch is often much easier to find than the bird itself in a new area. Kingfishers are territorial birds, and once you've found its perch, it is only a waiting game before you see the bird itself. Kingfishers are another sign of a river in fine health. If you suspect a kingfisher lives in the area, check the riverbanks for a single hole, about the size of a golf ball. Bank swallows dig similar-sized holes in sandy banks, but are more sociable, so there will be lots of holes in one area.

Signs that otters are busy in an area are easier to find than the nocturnal animal itself. Although otters are enjoying something of a renaissance after a bleak twentieth century, it is a good idea to research whether the river you are on is actually home to any otters, as they are still rare enough to disappoint otherwise. If they are around, then look for the otters' feces, or spraint as it is known, under bridges and on very low boughs or roots by the water's edge. Otters are sneaky; they love to swim downstream, but take shortcuts upstream, so if you're up for a bit of tracking, you can often find their cut-throughs over land where a stream goes round a bend. Keep an ear open if you are out at night and hoping to spot an otter, as they will make plenty of noise at times, mainly "eek" sounds, especially if there are pups around.

Otter country or not, it is always worth checking the water's edge for places where the grass has been smoothed down, as wet animals have a habit of "ironing" this waterside grass. Tracks in the mud nearby will then reveal the creature responsible— many are surprised to find that badgers are often the true culprit, not otters or other aquatic animals. Badgers are very happy to get in the water if there is food for the taking; they have been known to swim across to islands and devour all the birds' eggs there, before swimming back again.

If you see a heron peering into the water, then watch its neck carefully. Herons strike with a trademark velocity that you will miss if you aren't expecting it, and the clue is in the way their neck coils slightly, creating a slight "S"-shaped curve.

## WATER FEATURES

Having looked at many of the signs in the water's environment, it is time to zero in on the water itself. It is a good idea to find a vantage point where you can see across—a decent-sized lowland river bridge is ideal, or a raised bank, or even a tree.

Look closely at the water flowing in the middle of the river and at the edges and notice how the water is flowing faster in the middle than at the sides. It is typical for the water at the sides of a river to be traveling at only a quarter of the speed of the water in the center. The river is being slowed by two things at its sides; when it comes into contact with banks it is slowed by friction, and it is also slowed by the shallowing at the sides.

It follows that to win consistently at Pooh Sticks (If you're not familiar, in Pooh Sticks each person drops a stick off a bridge on the upstream side, and the person whose stick first shows on the downstream side wins), the simplest tactic is that you should always aim to drop your stick as close to the center of the river as possible. When my boys were very young, I made sure they were closer to the middle of the bridge, but they will be teenagers in a few years and it may soon be time to reclaim that berth secretly. I am a Romantic at heart, so don't enjoy the idea that something like Pooh Sticks tactics can be reduced to such a simple notion. Fortunately there is a complexity beneath the simplicity, as is so often the case in nature, one that water readers and Pooh Stick professionals ought to be aware of. It lies behind a little known, but very beautiful word: the "thalweg."

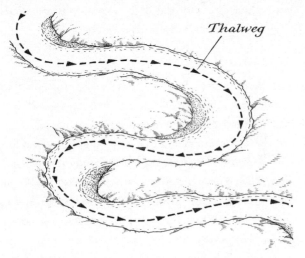

*Thalweg*

The thalweg is the line of greatest depth in a valley, whether or not there is a river flowing at the bottom of it. It is used by disparate groups, including lawyers and hydrologists; the former to solve boundary disputes (the thalweg is sometimes the legal limit of property, when two territories meet at a river valley)—and the latter to describe the line of deepest, fastest, and therefore most erosive water in a river.

The thalweg line itself is, of course, underwater and invisible if there is water flowing, but the interesting thing is that although the thalweg will tend to be near the center of a river, it will rarely be exactly in the middle, as rivers are never perfectly straight. In fact it meanders a little toward the outer bank if there is even the slightest bend in the river, which there usually is. This is vital knowledge for those who race boats of any kind on water, but without racing it is still well worth trying to spot the thalweg by looking for the slight variation in speed across the center of what looks like a fairly straight river.

## AN ISLAND IS BORN

As you study the river's flow, you may see the speed of water change for reasons that are not immediately obvious. If you do, then bear in mind that the water will slow as it goes over any shallows and accelerate as it goes over any dip. This can lead to a strange chicken-and-egg situation. More sediment will fall out of the water whenever it slows, so it follows that shallow mud banks grow as the water slows as it goes over them and more silt settles. Equally, when the water speeds up as it goes over any dips in the riverbed, it will deposit less and actually erode more, gouging out the dip. And so dips in the riverbed grow into bigger hollows and small mud banks in the stream become big mud banks, which eventually grow into islands. But what came first, the slow water that deposited the mud or the mud bank that slowed the water? Hard to say sometimes, but still nice to understand why a little patch of mud in the water you know seems to grow and grow.

In the slowest patches of river, where the water almost comes to a halt, the very finest silt is deposited. You may have experienced this if you have ever stepped into a river for a swim, as you are likely to have picked a bankside spot with very slow water. Here the line between land and water becomes blurred as your toes sink through the finest silt, barely slowing until your foot reaches the firmer mud below. There is a wonderful word for these patches of superfine silt in the slowest parts of rivers: the "cowbelly," as the belly is the softest part of a cow. Once an island has been born it starts to really dictate the pace, dividing the stream in two, each new stream behaving similarly to a normal stream, with faster water in the middle of each new stream and slower water next to the island and the outer banks. A single island in a river, or even a rock in a stream, will therefore create two fast streams surrounded by a total of four slower ones. If the river has not grown wider around an island than the island taking up room that the water used to have, it is sort of squeezing it. Watch the speed of the streams either side of it carefully and you will notice that the water in these is actually faster than the main river before and after the island. The island is behaving like a thumb over the tap, constricting and accelerating the flow.

*EDDIES*

Whenever a fluid flows past something that slows it down it is set into a spin, its whirls leading to whorls. This is as true for air and other gases as it is for water. Watch mist or smoke blown over a house and you'll spot how quickly it gets whipped into circular motion.

When water that is flowing down a river passes any obstruction—it might be huge like a bridge pillar or much smaller like a low twig from a tree—the water will begin to spin just downstream of that obstruction. This is an eddy. Amazingly and wonderfully, the physics behind this behavior of fluids is one of the most complex areas of science. In 1932 the physicist Horace Lamb summed this up amusingly with the following remark during a talk he was giving: "I am an old man now, and when I die and go to Heaven there are two matters on which I hope for enlightenment. One is quantum electrodynamics, and the other is the turbulent motion of fluids. About the former I am rather optimistic." And things have not grown much less complex in the decades since Lamb's words. But one simple truth is that the pattern you will see in these eddies will be totally unique. You will not see the exact pattern repeated anywhere else ever again, so they are worth a look for that reason alone.

Since the water in an eddy is flowing in a circle, it follows that some of it is actually flowing in the opposite direction of the river itself. Quite often a series of eddies are created that team up and lead to a steady flow of water in the opposite direction to the river itself. This has the effect of creating a very small stream flowing steadily in the opposite direction to the main river. This counter-flow is most common and easiest to see at the edge of the river where the riverbank is slowing the flow of the river's water. These counter-flows are always much thinner than the parent river and their speed can never exceed the main river.

Confusingly, the word "eddy" is used to refer to both the circular, swirling, vortex effect created as fluids pass obstacles, and also the counter-currents they set up that flow against the main stream. If you look along the edges of rivers you are bound to spot these little streams flowing gently the wrong way up the river. If you are struggling to find one, then the thing to look out for is flotsam, however tiny, floating the wrong way, as our eyes can pick this out much more easily than the water itself.

Even if you don't see the eddies close up, you can notice their effect from afar. If you look along a river from a bridge, notice how the sides of the river appear gently "ruffled" from a distance, and since no two riverbanks are identical, each side of the river will appear ruffled in its own unique way. My local river, the Arun, runs through the town of Arundel, and on one side the banks are built up for houses and made of smooth concrete and steel, while on the opposite bank there is a more natural mix of mud and grasses. Where the river runs past the natural bank this ruffling effect, the collection of lots of small eddies, is very noticeable, much more so than by the opposite bank.

The more powerful a river, the more substantial the eddies can grow, and where there are raging torrents there are eddies that are fierce and proud enough to have earned their own reputations and even names, like "Granite Eddy" in the Grand Canyon. Where a river is flowing fast one way and its eddies are flowing strongly the other, there is a turbulent line that marks the split between these two flows, known as the "eddy fence." On quiet rivers the eddies are just pretty and the fence hard to spot, but as the flow grows stronger those who spend time on this turbulent water, like white-water kayakers, come to know and fear these fences. They talk of "breaking" into or out of an eddy. In the words of one kayak guide, Rebecca Lawton:

> On the Colorado, eddies reign supreme. They're fierce, enormous, greedy, they could suck the *Queen Elizabeth* off course. Currents on their eddy fences boil so high, you need a stepladder to see over them.

This is not mere figurative speech. In the late eighteenth century, the Spanish naval schooner *Sutil* was caught in a great eddy near Vancouver Island and the sailors aboard reported the whole vessel being spun round three times, leaving them giddy. There are many who have experienced eddies but never lived to tell the tale.

But we don't need these savage forces or scale to find, witness, and savor an eddy. They can be found in the feeblest streams and are just as beautiful. Leonardo da Vinci was entranced by these smaller eddies and likened them to the flowing curls in a woman's braided hair. Look closely at any eddy and you will spot that it creates its own small eddies, too. In the 1920s the English polymath Lewis Fry Richardson appreciated this and celebrated it with a Swiftian ditty:

> *Big whirls have little whirls*
> *That feed on their velocity,*
> *And little whirls have lesser whirls*
> *And so on to viscosity.*

There is a slightly different form of eddy that forms underwater whenever water passes an obstruction or passes through a gap. This creates a whirling vortex of water under the surface that you can't initially see. After a short time, though, these vortices often do return to the surface, and they create a telltale upward-flow effect, a local upwelling of water. Sometimes, if you look just downstream from a bridge over a river, you will be able to clearly see these whorls appear back at the surface. They are often clearest when they appear to be totally redrawing a calm patch—it can look like someone has turned on a jet that points upward at the bottom of the river.

If the water is flowing very gently, this effect is often subtle and so forms the sort of gentle, rippling bulging of the surface that is only visible from certain angles and when the light is just right. This is a great example of the sort of effect that you might spot on big, quick rivers first, then on more modest, slow ones and then maybe, months later, looking toward a sunset as you lie by the weakest of streams, you may spot the faintest of little patterns and recognize an old friend.

## RIFFLES, POOLS, GLIDES, AND OTHER DELIGHTS

The river you are looking at will not be straight for long. Rivers do not run straight for more than ten times their own width, which means that if you find one that does you are looking at evidence of human tinkering. Canals run straight for much longer, but they are artificial and are best thought of as long, thin ponds, as both the banks and the way the water behaves have more in common with a pond than a natural river.

It follows that the broader a river the longer it can run straight for, but even the broadest of rivers will bend; when it does, interesting things start happening. Earlier we looked at how the water runs faster on the outside bend than the inside one (and the thalweg is closer to this bend, too). Fast water will erode and slow water deposits sediment, which means that a meandering river is far from a settled shape; it is in constant flux. Each day particles will be plucked from the outside bends and will settle on the inner ones further downstream, but over longer periods, the meanders in a river themselves will actually travel downstream. You can sometimes see this in aerial photos of meandering rivers or by comparing old maps with recent ones. And we all probably remember from school geography how the neck of a meander is sometimes broken by the river, creating an oxbow lake.

It is often hard to imagine a gently flowing river carving out the hard land, but imagine a cube of water as tall and deep as the average person and it will weigh almost three tons, so it doesn't have to travel very quickly to do a lot of damage. The difference between erosion on the outer bend and sediment deposits on the inner bend leads to them having different shapes and profiles, too. The outer bend will tend toward a vertical, small, cliff-like bank, which retreats over time, whereas the inner bend forms a shallow bar of gravel and sand or mud that grows over time.

IF THE WATER is flowing fast enough in a meandering river or stream, you don't get the wide sweeping meanders of the lowlands and school textbooks, but you still get some intriguing patterns that repeat. I find these patterns more fascinating than the more famous meanders, because they are everywhere to be found, especially in upland rivers, and yet, if you don't know to look for them, they are inconspicuous, becoming one more invisible component of "pretty scenery."

When water flows quickly enough through gravelly areas, it will carry the gravel a certain distance before depositing it. In places where lots of gravel is deposited, a natural barrier is created that then deflects the river or stream. The interesting thing is that this happens according to certain definite rhythms. There will be an alternating combination of quick water and much slower water, and this always happens in a certain way. The quick patches are known, perhaps onomatopoetically, as "riffles," and the slower areas are known as pools.

*A riffle and pool.*

If there is no human tinkering with a river's flow, then there will be a riffle–pool sequence for every stretch of river that is five times its width. So if you walk for a hundred yards along a river that is ten yards wide, you should expect to spot two riffle-pool combinations. Like so many of the water features we have looked at, the satisfying thing is that these features don't only appear on impressive rivers. The tiniest streams will show them. In fact, they display many more of them over the same distance.

Riffles are easy to identify because this is where the stream is steepest and the shallow water breaks into foaming white water over the rocks, making plenty of noise as it does so. Pools are just as easy to spot as they are deep, much slower, calmer areas of water, usually very near the outside of meander bends. In between riffles and pools you get glides, which are usually just downstream of pools; glides flow at a speed between pools and riffles and have a smooth surface.

These features are fun to spot as you walk past quick, gravelly rivers, but completely critical if you want to know what's going on in the river. And, as we will see in the next chapter, they are a matter of life and death, not for the water reader, but for the life within the water.

When kayakers approach rapids they have to be finely tuned to the features that are dangerous as well as those that are fun. Their insight into what happens to water when it hits huge boulders is helpful to us, even when reading the patterns around tiny pebbles in a stream.

One of the simplest of these water features to understand and spot is nicknamed the "pillow." Whenever a strong flow of water comes up against a rock or other obstruction in the stream, like the pillar of a bridge for example, it creates a bulge on the upstream side of that barrier. Like so many things we will see in moving water, the pillow is both stationary and fluid. The water is changing every second, but although the shape of the pillow may change slightly, it will remain in a fairly constant form for as long as the flow that created it does. Those traveling down the biggest and most serious white-water rapids in kayaks will see pillows as small hills of water that signal a large rock, but for most of us we can look for them as small, shimmering bumps on the upstream sides of more modest rocks and pebbles. It is fun to watch a leaf surf up one and imagine the ride it is on.

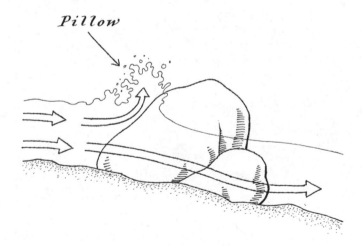

*Pillow*

There is a feature that kayakers rightly learn to fear: the "hole." When water flows over an underwater ledge and then drops suddenly, it speeds up and falls down to a point that is now lower than the surrounding water, which effectively creates a temporary "hole" in the water. Holes are both dangerous and interesting because of what happens next. The water tries to return to level and so water from around the hole will pour back in to try to fill it, and the slightly odd thing is that it will even flow upstream to fill it. Since the water is still pouring down over the ledge, the result is a curious and precarious balance where the water flowing down keeps creating the hole and the water flowing back up to fill it creates a continuous flow of water upstream, which in turn often results in a stationary wave that appears to be going the wrong way. The wave doesn't move upstream, but it appears to surge upstream without moving.

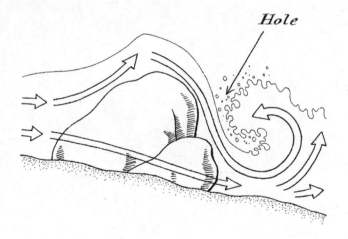

*Hole*

You can imagine that one of these holes on a big scale is something to be very wary of in a small craft, and getting caught, sandwiched between the two flows, is every kayaker's nightmare—they have killed many people over the years. But once we know what to look for, we can revel in seeing them on a much smaller scale. Some of the most mesmerizing ones I've seen were smaller than my hand. Holes are such important features of white water that they have many other nicknames, including "hydraulics" and "stoppers," but I prefer "holes" as it reminds us of what is going on beneath the surface.

After a while, features such as pillows and holes start to feel more like familiar animals than objects. To me, an hour looking for them feels like time spent stalking wild creatures.

# 6

# The Rise

DR. SAMUEL JOHNSON ONCE SAID, "Angling is an amusement with a stick and string: a worm at one end and a fool at the other." Sir Humphry Davy countered that it was more a case of "a fly at one end and a philosopher at the other." I traveled to the Peak District, the upland region at the southern end of the Pennines, already confident who was right, but with a plan to make some inquiries of my own.

Stuart Crofts shook my hand when we met in the village of Castleton, in the Peak District, and before he let go he was assuring me, in a thick Yorkshire accent, that we would soon be discussing with the river the things it would be happy to tell us.

Stuart describes himself as one-third fisherman, one-third entomologist, and one-third overtaken by a childlike enthusiasm for all nature. I had arranged to spend the day with him to help me fine-tune my reading of water in one specific area. I am no fisherman or hunter and, if I am honest, I have never had any great desire to be either, but I have long respected the deep wisdom that both hunters and anglers develop for their niche in the natural world. It is a wisdom that often brings a calm confidence outdoors, and one that allows a little self-deprecation, too. Stuart laughed, recounting the ribbing that his young daughter had given him when he had tried to impress her with a catch: "Congratulations, you've just fooled a creature with a brain the size of a pea."

It is only possible to grasp the artistry and passion of fly-fishermen when we appreciate that the catching of the fish is a very minor part of the appeal. I asked Stuart, a man who has dedicated his every possible waking moment to the sport and the nature around it, how he would feel if he was told he would never be allowed to catch another fish in his life.

"Wouldn't bother me in the slightest," he replied calmly and sincerely, and I had absolutely no reason to doubt him; I get it. Dry fly-fishing may date back to the Macedonians at around the time of Christ, but it was the Victorians, who embraced it as a pastime and this was the moment it began the leap from food for the stomach to nectar for the mind. Brian Clarke, a recent luminary in the field, put it very well: "The key to it all is thought . . . The expert thinks more in terms of how and why, than of what." Clarke is convinced that it is not about the tackle you have or the techniques you employ, but about your understanding of your environment. Because fly-fishing is about understanding the water, the fish, the insects they eat, and recognizing how the slightest breeze or even the movement of a cloud across the sun changes everything.

The idea that a tiny change in our surroundings has a broader impact is commonly talked about, but in fly-fishing you see it actually happen. Flying insects live on the edge of death every second of their short lives; the very fact they are flying at all is a precarious balance, dependent on how hydrated they are (most insects die from dehydration) and factors like how warm they are. When the sun slips behind a cloud, insects will cool slightly, and some lose the ability to fly and drop out of the air onto the river, where a trout will be expecting them. It is this sensitivity that makes the angler.

"Very little in fishing is down to luck," Stuart told me, and when I laughed because I thought he was joking, he said he meant it. He has a fine sense of humor, but he wasn't going to let laughter dilute the important stuff.

There was no arrogance in Stuart's learning; he spoke fondly of all, including those who "can't tell a bumblebee from a bull's foot!" Like everyone with a true love of the outdoors, Stuart is a model of sensitivity to his environment: He is finely tuned not only to what is happening around him, but also to the impact he is having. We would be traveling downstream over the course of the day to ensure biosecurity, so that any organisms that we inadvertently gave a lift to would be taken in a harmless direction. In sensitive ecosystems, which of course is all of them, walking upstream and getting in and out of the water runs the risk of allowing hostile infestations to piggyback into previously unsullied waters. Heading downstream offers no favors to rampant invaders like yellow star thistle and zebra mussels.

Stepping down from a road, through a patch of wood sorrel, our boots crunched onto a forest carpet of needles. We had only taken a few steps alongside a tiny stream that trickled through the dark soil, when Stuart thrust his hand down into the water.

He used his fingers to churn up the silt and then we waited for it to settle. There, in a place I would never have considered looking for it, was a teeming mass of life. However much I try to remind myself that I am forever overlooking insects, I continue to underestimate the richness of their world around me. It is possible to find an insect unknown to humankind in your own back garden, literally; it has been done and recently. Imagine that, getting an insect named after you, because you were intrepid enough to rummage around a few meters from home, how wonderful!

The water ran away downhill and carried the disturbed silt with it, leaving a small, bare patch of tiny gravel stones. Wriggling around on top of them were dozens of freshwater shrimps. A few seconds later we were inspecting cranefly larvae and a caddisfly.

"These are wonderful, but what do they mean?" I asked Stuart. I had warned him that my curiosity was driven by understanding the clues, the signs, the patterns. The beauty of all organisms only really flourishes for me when I can understand what they are trying to tell me.

"Freshwater shrimps are a good sign: They mean we've definitely got very low ammonia levels in this water, as they won't tolerate that at all. So we've got no human or animal waste feeding into the water anywhere upstream. This collection of insects here also confirms that we've got a slow-moving stream, slow enough for it to be silty." Those two things are connected and very important to insects and consequently fish, as slow, silty environments are very different habitats from faster, bare ones. We gazed down the stream and as we did so, we discussed how so many of us are keen for our local waters to be as untainted as possible, but then we mostly depend on the government and other third parties to give us the information about whether they are pristine or not. Occasionally people will note that salmon have returned to a river, but salmon are late markers of what's going on. If we have concerns about what is coming out of a pipe that feeds into a river, we need only to take an interest in the aquatic insects and we will be able to make our own reports. By noticing the different insects we find when comparing the water upstream and below the pipe, it becomes impossible for politicians, businesses, or anyone else to mask the truth.

Stuart pointed to a collection of small insects on a small, flat rock, which had built tiny, kiln-like houses in the stream, and he explained that the Agapetus (*Agapetus fuscipes*) he could see there needed a very high water quality consistently for at least one year, so it was a sign that there hadn't been any pollution of the water for a single day in the past year. They were also a sign that the water was dependable and the spring was unlikely to dry up in the summer, as they couldn't cope with that. Other insects with longer life cycles indicate that the water has held its purity and flow every day over two or three years.

INSECTS ARE one of nature's more ingenious ways of making us sense the passing of time. Anyone who has ever visited the Scottish Highlands is likely to have discussed the best time to go to avoid the worst of the midges there, but when it comes to understanding the relationship between insect and fish behavior, a new level of awareness is called for. Take the mayfly, one of a trout's favorite meals. It lives in the mud for two years and then takes to the air for a single day—without a stomach, as it literally doesn't have time for one. For trout fishermen, understanding when that day will fall each year is filled with anticipation, and so the clues to any insect hatch form a vital part of their puzzle.

"Wagtails are a sure sign that a hatch of fly is about to emerge," Stuart told me. "So are blackheaded gulls that appear from nowhere on larger rivers. A thinking angler will head straight toward this activity because if the gulls are taking the emerging insects, so will the fish."

From the catkins on the willows that form an early source of pollen and nectar for the insects, to the tiniest shift in water temperature, the rivers, streams, and their banks are poised on the brink of a series of explosions into flying life from spring to autumn. Just as some wildflowers, like St. John's wort, earn their names from their coincidence with seasonal dates, so the angler's insects will earn colorful and useful monikers; the St. Mark's fly hatches near St. Mark's Day, on April 25th. This is one of the quickest ways to discern the level of a fisherman: Ask him or her about the insects. There are catch-all expressions to save those who struggle; "olives" is a term used for a broad range of insects (just as birdwatchers rely on "LBJs"—little brown jobs!), but the real experts will be getting up close and personal, usually with some magnification.

Stuart and I had descended a little way and emerged from the dark conifers, past the curdling hollers of a concerned shepherd, and came to rest by a broader stream that sparkled under the bright sun. An orange-tip butterfly came to investigate us but found little of interest and continued its own journey. Net in hand, Stuart stepped out into the stream and soon new friends were being decanted into a white tray for me to investigate.

"How many tails?" Stuart asked.

"Err . . . three," I replied.

This was to prove the question that helped make sense of, and bring some easy order to, what is potentially a daunting part of the animal kingdom. If the insect had three tails it was part of the mayfly nymph group, also known as the Ephemeroptera or "upwings." If it had only two tails it was one of the thirty-four species of stonefly nymphs. Up close, the mayflies move with a dolphin action and the stoneflies with a crocodile one.

If you are delving into the world of aquatic insects you will come across the words "nymphs" and "larvae" being used to refer to the immature stage of these insects. It is worth knowing that these are not references to two different stages in the same creatures, but a way of differentiating between the insects that metamorphose as they reach adult stage and those that do not. Nymphs are insects that sprout wings and become airborne, but don't metamorphose, whereas larvae will metamorphose into a new form. But beware the textbooks that use these terms interchangeably and lazily!

The three-tailed individual that I was admiring was a Stone Clinger mayfly nymph; it earned its name by its habit of using the pressure of the water flow over its head to stick to rocks, and it is a sign of the finest water quality possible. Stuart explained how, with enough knowledge it would be possible to analyze every possible input, pollutant, and stressor to the water from the insects found there. Nitrates, phosphates, oxygen levels, light levels, water speed, predators, each individual pollutant . . . there would be clues to all of these, not just at that moment, but their levels over every second of the past year or more.

Stuart searched patiently to find me a very rare local speciality called the Upland mayfly (*Ameletus inopinatus*) and explained that it was of great interest to local entomologists as they were using it as their canary in the coal mine, the insect that is most sensitive to climate change. Then, returning to a level I could master, he illustrated how, in a tray full of insects, there was not a single freshwater shrimp—the water was now running too fast for them.

Stuart returned our friends to their own water, and as he did so I stepped forward and, looking at the water in the direction of the sun, I marveled at the beautiful way the white water around a rock was effervescing, hundreds of tiny diamonds being thrown up into the sun each second. But I had no idea that the insects around me were mapping this effect, too. Stuart explained that the insects' vital need for moisture meant that they were very sensitive to the humidity of the air. The bubbling parts of rivers and streams created a layer of much more humid air above them than the slightly calmer water that was very nearby. This meant that the insects would gravitate toward these patches of white water, drawn by the moisture in the air. Would we ever have discovered this without the passionate inquiry of the anglers? I'm not so sure.

The insects have another trick: They can identify polarized light, and all light that reflects off the water is polarized. To an insect, light reflected off water looks totally different from light that comes direct from the sun. (If you own a pair of polarized sunglasses you can get the tiniest hint of how this water looks different to insects by tilting the glasses on your nose and observing how patches of water change in appearance slightly. The debate about whether sunglasses help when observing water in general rumbles on; they reduce glare, but reduce all other light, too. Personally I prefer not to use them on land, but will on sunny days at sea. With or without sunglasses, a good general tip is to look at shaded areas first, before turning to brighter ones, as this gives your pupils time to adjust more comfortably and effectively.)

"It's a Tiger Moth!" I said, pointing like a child, as I caught sight of a flying insect in the sun that reminded me of the biplane.

"What did you say?" Stuart asked. He sounded stern and I thought perhaps I had said something wrong. A second later, he asked again, but this time I could tell he was excited, not alarmed.

"That insect looks like a biplane. It's got two sets of wings."

"Ha ha! That's brilliant!" he said, which surprised me as I couldn't see how that was possible.

"It is?"

"Yes. I call it the Sopwith Camel, but same difference. Look for the Sopwith Camel, I always say. It's the Stonefly."

I laughed and looked around me, noticing how much easier it was to spot both the white sparks of water flying off the stream and the insects in the air above them by looking in the rough direction of the sun. We sat down for a bit and I sipped from my flask of water as Stuart explained that the serious angler should net insects both in the air and in the water upstream of where they plan to cast. He showed me the different nets he used for that purpose. And so it was that I learned how the person who heads out with a rod thinking they are only interested in catching a fish becomes an entomologist by accident.

Stuart and I discussed how the elements and insects interact in a way that passes so many by. We looked at the black gnats by the stream and Stuart explained that they were a good example of a flying insect so sensitive to temperature that if they are flying over water when the sun slips behind a cloud, they will drop out of the air and into the water.

If we combine this sensitivity with the curves in a river's path and the wind direction, even each individual shift in breeze, it explains why there will be a rich collection of insects in one patch of water and none a few yards away. And the fish are perfectly tuned to these differences. This is why there will be one set of anglers with smiles, while their neighbors around the bend in the river are growling and blaming their kit. Stuart's words about luck came back to me, but he added a few apposite ones as he half-growled, "It's them bloody signals you've got to tap into!"

We walked along the broad stream's banks, Stuart pointing occasionally at spots in the water: "10 percent," "30 percent," "10 percent," "70 percent," "ooooooh, 100 percent, there is definitely a fish in there." We were looking at "pockets," the small patches of calm water sitting just outside the main flow as Stuart assessed the likelihood of there being a fish in them. He paused and pointed at a water form I recognized only too well, "Look at that eddy. Got to be a fish there, bloody got to be!"

Stuart never pointed at the rough water of the riffles, or at the very smooth water of the "glides," always at the calm pockets to the side of the white water. These pockets are the same feature, just junior versions of pools on a bigger river. The topology of a river, the very same detail that can appear flat in geography text books, becomes rich and deep when we realize that the sequence of riffles, glides, eddies, pools and pockets are a map of the life in the water. The fish are constantly seeking the best deal, the most food for the least effort (as all creatures on the precious edge of starvation always are; small birds live in this state almost permanently).

At the same time, the fish are doing their best to eat without getting eaten, and all fish that reach maturity have learned this lesson well—not only must they avoid becoming food for birds and mammals, but also other fish: All fish will eat other fish. This means that the physical struggle to feed in the fast, shallow water is to be avoided and they cannot afford to wallow in the slow, perfectly clear water of the glides, where any hungry bird will spot them. They must be well tucked away, among the roots and out of sight until nighttime. The pockets that sit just to the side of flowing water bring food in the form of insects past them like a conveyor belt, and if these pockets are sheltered and shaded by some features like gnarly tree roots or conveniently shaped boulders, even better. Stuart was assessing each of these factors as we followed the stream, each one nudging his prediction of the probability of a fish living there up or down a bit.

"They love the popply water," he said.

"Popply water?" I said, worried that this may be a formal term I was totally unfamiliar with.

"Yes. They love the pockets and the right pools, but if they come into the main stream, it will be in the popply water." He showed me what he meant. In the riffles, the water is mixing with air and that is what we hear so well. In the glides, it's all calm, but in between you get the "popply" water, where the water is running over the rocks, but not with enough speed or energy to break and mix with air. "It's neither flat nor white and mixing with air, just . . . well, popply. The fish love it." I recognized the type of water he meant. It has its counterpart in the sea, as we shall see. It is worth repeating that it's the breaking of the water, this mixing with air, that creates the sound in running water, so we hear riffles but glides, pools, pockets, and popply waters are silent—you have to spot them.

The fish also love the "soft spots" and "friction zones" around certain rocks. The water just upstream and downstream of a prominent rock will move slower than the main stream slightly further from the rock. These are the "soft spots." And it will often be slower on either side, too. These are the "friction zones." The fish love both.

Stuart's finger darted out regularly toward a patch of calmer water on the other side. I asked him why he was only pointing at the pockets on the far side of the river; surely there were some good spots on our side? He stopped walking and grinned.

"Ah." He looked both excited and sheepish. "Well, yes, you're right. It's because I'm right-handed." He made a casting gesture with his right arm and pointed at the far bank. "Right-handers will see the river differently from left-handers. If I walk a river with a left-hander we will zero in on different spots. Sometimes I trick myself into believing that I'm going to cast with my left hand, and this reveals pockets that I would never have spotted otherwise."

I loved this idea and shared a similar one that affects walkers. As we pass tall obstacles on a path, we will use our less-favored hand to "fend off" that obstacle. Imagine you are a bit lost and on a path that forks either side of a tall thin rock. It would be easy to imagine that you might choose either side of that rock, but that is not true, as we all have preferences loosely programmed into us, and these are influenced by whether we are right- or left-handed. Right-handers like to put their left hand on the obstacle, probably because this keeps their favored hand free. It can lead you to walking in circles if you're lost in rocky terrain and are not aware you are doing it.

A lizard crossed our path, and then paused to assess us and soak up some more April sunshine. A little further on we stepped over the drying remains of a sheep's afterbirth as Stuart explained that he loved to ask children what the oldest extant living thing is. "They usually end up on something they think is like a dinosaur. I tell them the mayfly was around 150 million years before the dinosaurs and they're still going strong." There was a pride in his voice, as though the mayflies were part of his team and together they had beaten evolution.

A patch of oil floated past on the water and we stared at the iridescent colors and considered its origin, as it drifted close to the sun's reflection. We both thought the resin in decaying pine needles upstream was a more likely culprit than anything industrial.

Stuart paused between a row of conifers and a rock ledge and we stared down into a calm pool, at the edge of some white water.

"The trout needs two things: shelter and food." He explained that if we can learn to see the river in terms of those two things, from the trout's perspective, then we will find the trout. Places that offer shelter in combination with water that isn't too fast, but also near enough water that is flowing and bringing a concentration of food past, will be prime real estate for the trout and guarantee their presence. We were looking at one of those spots.

"There!" The finger shot out again. "Did you see it?"

"No," I said, straining my focus, following Stuart's finger as best I could down into the water on the far side.

"There, again!"

"Yes! I saw it!"

The ripples spread out, leaving a calm center, before dissolving into the more turbulent water around them. I was ecstatic. I did not want to put unfair pressure on Stuart, but this had been my big hope for the day. I could not have been happier than a Victorian big-game hunter that had taken down a beast on the African plains. I had seen "the rise." For someone without a desire to actually catch the fish, this was a spectacular triumph and moment. The day up to that point had been us stalking this moment, studying the weather, the water, the plants, the birds, the insects . . . all of it allowed us to creep up on this wonderful sight of the trout rising, showing itself at the water's surface with a signature pattern of ripples.

LIKE ANY ART, fly-fishing is not immune to passionate debate about the finer points. But the beauty for me is that there is an art in fishing without fishing. Let us call it "Rise-Watching."

And the same debate can surround it, because it is a rich and rewarding one. Fly-fishers adore to see fish rise, even if they end up not catching any. The rises are what demonstrate the activity to the rise-watcher and the potential to the fly-fisher, and are equally exciting to each. The rise-watcher can take their lead from the excitement of the fly-fisher and chalk-water conservationist Simon Cooper:

> That moment when a dimple radiates across the river surface just ahead of you as a trout sucks down a fly is truly every bit as good as eyes meeting across a crowded room.

No two experts will agree on the exact form that arises when a fish breaks or even nears the surface. But the areas of agreement are tied by the logic of fish feeding behavior.

The fish, and we will focus on trout for simplicity, come to the surface to take an insect. We know how sensitive water is to the slightest change around it, so when any fish takes an insect in their mouth, they cannot do it without disturbing the water's surface. This creates the rise that we can spot; so far, so simple. But what exactly will this rise look like and why? And what can we deduce from the subtly different rise patterns that we see? These are the questions at the heart of fly-fishing and rise-watching.

There are a few fundamental principles that all are agreed on. The trout's food, the insects, are varied in form and behavior. There are small ones and large ones, ones that fall dead from the sky, ones that are trapped and wriggling, and ones on the surface that are ready to fly away an instant later. Imagine a trout spots a very small, motionless, probably dead insect on the water's surface. It is neither much of a meal nor likely to escape in a hurry, and so the trout is not going to waste much energy on catching that meal, it will approach leisurely and eat gently—why waste lots of energy on a surging dart and energetic snap? But a much larger insect that is very much alive and poised to escape, that is a different challenge for the trout—a proper meal, but one intent on not becoming one. And so the trout will approach this much more like a smash-and-grab raid.

The various strategies employed by fish to take their meal at the surface lead to the wide variety of rises that anglers subscribe to and look for. Depending on the authority you refer to, there may be kissing, sucking, sipping, slashing, flushing, kidney-shaped, or bulging rises. The debate and disagreement, even between long-hallowed experts, about the exact form that these rises take is bewildering. But with Stuart's help I will shortly try to simplify this area.

Below the trout's rises, even the subtlest kind, there is still plenty of spotting to be done, of course. Sometimes a fish that is swimming just below the surface will disturb the water in a way that is so subtle that it wouldn't qualify as a rise, all but invisible to most observers. But it will be noticeable if you are looking at the right sort of reflections. The straight, clear edge of a tree trunk may become blurred or may flex slightly, perhaps even twisting in a full "S" shape. (It might help at this point to think back to our work with the "Seismometer puddles" in the Puddle chapter.)

Trout will change color rapidly if necessary, in a matter of days, to suit their environment, and they are masters at being inconspicuous to bird's eyes, let alone our own weaker instruments. They are so good at changing their appearance that the Victorians classified several different species of brown trout, when they were looking at only one species in many different clothes. But we are not powerless and trout will betray their whereabouts when hunting for nymphs in the following way. They face upstream, occasionally shifting their position to the left or right, before returning to their original position, all of which might be hard to spot at first until the telltale sign reveals itself: the *chink of light*, which is the small patch of white that appears against a dark background each time the fish opens its mouth.

You're unlikely to spot the fish's tail itself straight away, but it's worth being alert for the rhythmic movement of the shadow of the tail. The best general rule is this: Search for any anomalous movement, because even though camouflaged animals can make themselves almost invisible, especially underwater, one weakness of camouflage is that it can't disguise movement very well, as the background doesn't change to keep up with the fish. Occasionally a fish will ruin its own camouflage by fanning silt away from a gravel patch below it, making it stand out against a lighter background.

Unsurprisingly, when looking for the fish themselves it helps to give some thought to the sun and wind. Calm days when the sun is high and behind the shoulders make looking into water easiest, but bear in mind that the fish will be sensitive to any breaking of their horizon by you. You can improve your odds by increasing the light you need and decreasing the light you don't, which essentially means screening out the sky with a wide-brimmed hat or cap.

Stuart and I had been watching the rises from the same spot when he ran me through the combination of factors. The insect behavior, the wind direction, the slow pool next to the faster water, the sunlight and shade on the water, the fact that we had a line of dark trees behind us so that we were not breaking the fishes' horizon. We watched a sequence of three rises, each one triggering a muted, excited reaction on my part. By watching the sequence of these rises, you can quickly work out if it is several different fish in the same pool, or the same fish on a circuit. If it is on a circuit, predicting the exact spot of the next rise gets easier.

"One, two, three . . . there! Same fish," he whispered, and we watched until the pattern was repeated. Then we edged uphill to gain a different perspective and the rises stopped. We had passed the edge of the trees and were now breaking the horizon. The trout were sensitive to our every movement now and had darted for shelter.

"People don't believe me when I tell them this, but it's true . . . When I fish at night, I cast by listening for the rise. Seriously." I believed him. "There, look at that scum lane."

"Scum lane?"

"Yep. Where the bubbles flow in a line down the river. It shows us where the forces of the water and wind are collecting things at the surface. That's where the insects will congregate. We'll see a fish there if we're patient." We didn't have to wait a minute before a set of concentric rings spread out and then another and another.

"It's not scum, as in dirt though, is it?" I didn't like the idea of any dirt in this purest of rivers.

"No, that's just the name for the bubbles, the bubbles just come from the white water, the riffle up there."

I watched for the next rise and spotted it easily, but then my mind drifted to all the different types. Having spent a long time wrestling with the different rise forms and failing to sift one from the other at all effectively, I put the matter to Stuart. He was very diplomatic—perhaps unwilling to speak blasphemously about the great names in his sport—and said that each person sees what each person sees and that it's not something you can be "wrong" about, so long as you're honest in what you report seeing.

He seemed to suggest that there was a subjectivity to the perception of the rise forms, which is fitting in light of considering this as an art form. Perhaps it was a question of the level of detail that each person wants to see, one person's splash being another's "double kidney shape." I pushed him and asked him which rise forms he personally identified and used. He paused to consider his answer, and my eyes were drawn to the swallows that were dipping down into the water below the bridge, for the briefest of drinks on the wing. He explained that after forty years of passionate fishing, he grouped the rises into three categories. I almost sighed. But then it became clear that there was only one in each of Stuart's categories and my hopes lifted.

"There's the kissing or sipping rise; think of a grandmother gently rocking in her chair. She asks for a teaspoon of gin and you have to touch it to her lips ever so gently. This is the kissing rise." It was the rise we had seen earlier.

"Then there's the splash: when the fish is moving at pace and its head often emerges . . . sometimes you can see its eye!

"Finally, subtlest of the lot, there's the subsurface. Very hard to spot, I sometimes call it 'nervous water.'" This was a rise I had referred to as "the bulge" by others. "When the fish takes something below the water without breaking the surface, although its tail sometimes tips up . . . No use fishing with a dry fly for a subsurface rise—you're wasting your time!"

We moved away from the river's edge through an air thick with wild garlic scent and between two wood anemone carpets. "It's a game of chess. But you might only get one move," Stuart said, as he unpacked a gas burner, kettle, and mugs from a wooden cube. We enjoyed a cup of tea, and I could not resist showing him how the lesser celandines and daisies in front of us were arranged to point south. Over tea the conversation became yet more philosophical, as Stuart talked about his approach more broadly; his desire to blend in so that the river doesn't know he's there. I was struck by the way he liked to use the word "river" as a short-hand to describe not just the water, as many might, but the intricate network and ecosystem of which the river is but one artery.

"It's about allowing the river to invite you in, so when you're walking up the river you may put your hand on a duck while it's still sat on its nest, or you get a kingfisher flashing past you, having to take avoidance action so it doesn't crash into you, or a dipper, or even a heron that gets up and you feel the waft of its wings . . . that's when you've been invited in t' river and . . . that's the point at which you start to become a true fisherman or a true hunter."

Until that moment comes, there is no harm in any of us pausing at a bridge and looking down for clues as to where the fish will prefer and then watching for a rise. If I was offered the choice of a freshly caught trout and the sight of a rise exactly where I have worked out one is likely, personally I'd go for the rise. It tastes less good, but serves up better memories.

# 7

## The Lake

THE WATER VOLE DISAPPEARED into the slow stream and the sound of its small splash was almost lost to the wood pigeons and chaffinches. Bright pyramids of pink-tinged white flowers covered the horse-chestnut trees and fair-weather cumulus clouds scudded by above them. Conditions early on this late spring morning were perfect for my plans, with plenty of life, plenty of light, and a lively breeze. I was walking at the edge of Mill Lane, toward Swanbourne Lake, the words sketching a picture of an area rich with water. The lake itself dates back to before the Domesday Book and was originally used as a mill pond to supply Arundel Castle in West Sussex, in southern England. The "bourne" part reveals that the lake is fed by springs, from water that has filtered down through the surrounding chalk. This is one of the first things to consider on arriving at any large pond or lake: What lies underneath?

The rocks beneath a lake will have a huge bearing on the plant and animal life both in and around the water. If you have walked across peat to find the water, you are in an acidic area, and there may be many dragonflies, but plant and animal life on the whole will be restricted. There will be almost no crustacean life, for example, as there is neither the calcium for these animals to build their shells, nor any way they could survive the ravages of acidic water. If you are in a chalk landscape, then the diversity of plants and animals shoots up and crustaceans are very likely to be found in good numbers.

Swanbourne Lake is and always was an artificial creation. It may be an old part of the landscape, but it would not exist without human landscaping. Artificial lakes lack charm for some purists, but they are in good company: In much of England, especially the south, very few lakes are "natural." Alongside the rock types, it is worth giving some thought to the broader geological history in any area, because if you are in a region that once had glaciers gouging and barging their way through it, then it is much more likely that the lakes would be there whether or not humans were. But south of the glaciers, ponds and lakes are very rare without some human assistance. There are a handful of exceptions, for tectonic activity can create some of the deepest and most interesting bodies of water, like Loch Ness or the great lakes in eastern Africa. And we mustn't forget the odd river-sculpted one, like oxbow lakes.

Once you've gauged that a pond or lake is likely a human artifice, you can ask what its purpose was, which will in turn reveal something further about the water and the area. A little mystery in ponds and lakes is no bad thing either, as they offer a puzzle to be solved. I came across a freshwater pond very close to the sea in a south-coast English town called Littlehampton. I wondered if it had been built for aesthetic purposes, but was not convinced, as Littlehampton is a town that does not appear to have been overwhelmed by such considerations to date. There was nothing to give away its use in its shape, but the mystery evaporated when I learned its name. The Oyster Pond was indeed once used for holding oysters.

Whatever the physical cause of freshwater accumulating in one spot, it is worth admiring, if for no other reason than it is rare. A walk along a broad river or a trip to somewhere like North America's Great Lakes region can trick us into believing that fresh water is plentiful, but this is far from true. Ocean water is abundant, but for every 6,750 liters of ocean water in the world, there is only one liter of fresh water in rivers or lakes. This is why its use is contested, even in relatively wet parts of the world.

As I arrived at the edge of the lake, it made its presence known. There was a familiar wet scent in the air, tinged with the weakest hints of bird droppings and damp, decaying foliage. By taking note of the smells around water we tune quickly into the wind direction, the nature of the shoreline, the vegetation, and the temperature. One sniff of the air near water can reveal a shift in the seasons. If all you pick up is the faintest of watery scents, then this suggests all is hopefully well, but it is normal for smells to be a bit stronger in summer than winter, as all life, and especially the algae and bacteria in the mud, is more active. However, the slightest whiff of rotten eggs indicates that the bacteria are producing hydrogen sulphide, which in turn means that oxygen levels are low and the water ecosystem is deteriorating.

Scent-mapping places is a good idea for another reason. The information that our brain receives from our sense of smell takes a different route from that taken by the other senses, bypassing an area called the thalamus. Smells reach a part of our brain associated with both memory and emotions more directly than sights or sounds. We form another impression and build another layer of understanding about a place when we register its smell, but we can form a longer-lasting and more emotional map of it, too. I won't trouble the reader with all the smells that bring back certain places for me with irresistible force, but there is one smell that still brings me close to tears after more than twenty years, and I've no doubt you have your own powerful examples. In an age of geo-caching, where walkers use GPS screens to find strategically hidden boxes, there is hopefully still room for us to enjoy some geo-sniffing.

There are many ways we can use our sense of smell practically, from appreciating that the coast is getting nearer to the lost desert navigator who found camp from the scent of a single camel five miles away. If we want to build an accurate picture of where we are and what surrounds us then, in the words of the nautical expert Tom Cunliffe, we shouldn't "ignore the faintest whiff of a rat."

Back on the shores of Swanbourne Lake, I accidentally woke some geese and gulls who looked unimpressed and waddled away, although how they managed to sleep through the perpetual racket that the squabbling coots were making was a mystery. I peered into the water.

Swanbourne is a shallow lake and it is what hydrologists would call "holomictic," which means that the water is shallow enough that it all mixes and is therefore all roughly the same temperature. Depth is such a fundamental feature of all water, but especially still water, because the depth determines the light and temperatures that will be found in the water, and these are what determine the life or lack of it. A few yards will make a massive difference, because light is absorbed very dramatically by the top layers. Of all the sunlight reaching the surface of a lake, perhaps less than half of that will reach one yard below the surface, only a fifth of the total will get down to two yards, and a tenth to three yards below the surface. The exact levels the light reaches down to will depend on how clear the water is, of course, but in any lake of depth there will be distinct zones, which can be thought of in terms of light and temperature.

The euphotic zone is the layer of water that light can reach, the top layer, and is where plants can theoretically grow. It ranges wildly from fifty yards deep in very clear lakes to only tweny inches in very silty waters or those thick with algae. Lakes also form zones according to temperature. In deep-enough lakes, there will be a bottom layer, called the hypolimnion, that holds a constant temperature near 39 degrees Fahrenheit. (It is not a coincidence that this is the temperature that water is at its densest. One of water's peculiarities is that it becomes less dense as it both warms up from 39 degrees and also when it turns to a solid, as ice. Hence ice floats, but cool water sinks below warm water.)

Near the surface, where the sun warms the water, there is a different layer called the epilimnion, where the temperature varies wildly over the course of the year, from near-freezing in winter to warm, swimmable temperatures in the summer. The two layers are divided by a layer called the thermocline, and its depth varies with changes in the wind and sunlight that gets in. These zones and their depths fluctuate with the seasons, which is why a lot of pond and lake life gravitates lower in winter and makes the water appear dormant, if not dead.

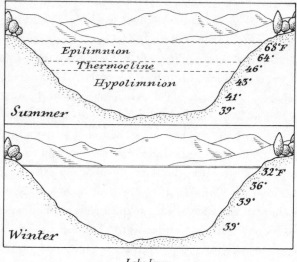

*Lake layers.*

You can actually create your own thermocline experiment in the kitchen if you want to. All you need to do is find a transparent rectangular container like a glass baking dish and fill it with about one and a half inches of cold tea. Now carefully pour three quarters of an inch of warm water over the back of a spoon to stop it mixing and you will be able to see a layer of warm liquid sitting on top of the cooler layer, the two layers divided by a thermocline.

In the oceans things become further stratified, with haloclines dividing zones by saltiness and with more zones with wondrous names like the abyssopelagic (2.5 to 3.7 miles deep). The oceans also have a thermocline, typically between 660 and 3,000 feet, which creates a sound barrier that blocks sonar—it is used by military submarines to hide from each other.

Incidentally, temperature layers in the water itself have a big impact on how sound travels underwater. Whales can sometimes communicate over vast distances, and some scientists believe they have evolved to use sound channels underwater. By bouncing their song off layers of cooler, denser water, this may be a key to whales staying in touch over thousands of miles.

These zones—with their Scrabble-busting names—make water depth and its relation to light and temperature seem very technical, but really they hide a basic truth. We only need to be aware that we should expect to see a reduction in plant life and animal activity with depth and that is not always gradual, as there can be dramatic temperature or light cut-offs.

Some of the most interesting clues to water depth will be found at the surface without any need to peer deep down. My walk around the lake took me past water lilies, which as we saw earlier, are one of nature's depth gauges. Stepping around a jutting field maple, I watched as the white rumps of rabbits disappeared up a bank of wood spurge and into the woods, before the path led me to a break in the woods and it was here that I saw perhaps the most elegant depth gauge of them all: a swan. Swans feed off the algae, aquatic weeds, and roots they find on lake bottoms, so they will be much more common on the shallower parts of a lake. The swan I saw had three cygnets with her, and they would have made a hungry gang, further improving this pretty depth gauge.

In truth every animal and plant you see on the water is giving you some idea of the water beneath them: If you see a duck that looks like it's lost something and is frantically searching in the water, then you're looking at a shoveler duck, as it stirs up the bed for food with its great broad beak—obviously this must be very shallow water. At the opposite end of the scale, cormorants have been known to dive to 150 feet.

I was now walking in full sunshine and enjoyed watching the reflected patterns dancing on the trees next to me. Water milfoil, which does better in shallower water than the lilies, had colonized a large patch of the lake in this corner. I traced it to near the edge and then watched as the plants segued from these true aquatics to the plants on the bank, like hemp agrimony, that are terrestrial but thrive in damp soil. Near the edges of the lake, dust, leaves, and other natural debris were collecting in pockets among some bulrushes, and in places it appeared to have congealed into something closer to mud.

This was a reminder that ponds and lakes are far from permanent; rivers will tend to grow naturally with time as they do their own excavating, but the opposite is true for still water. Unless ponds and lakes are given some help, they will all eventually fill in and return to land. It starts with algae, then the rushes and other shallow water plants getting a foothold, and this allows sediments to gather, water turns to wet mud, and a reinforcing cycle begins that culminates in the water losing the battle against the encroaching land.

The path rose up a small hill and this gave me a better vantage point and an opportunity to study the surface of the water. The better view allowed me to enjoy studying the finer points of the breeze and the ripples.

WHEN THE WIND is blocked by obstacles, calm patches will be found in the water downwind, but there is a fine art to reading these patches and there are some subtleties worth looking for. We can thank sailing strategy experts for a forensic understanding of how breezes behave around obstacles, because this knowledge can give racing sailors the competitive advantage needed for victory.

When a wind has to pass over an obstacle, it will not get back to its former strength and character until it has traveled a distance that is approximately thirty times the height of the obstacle.

But there are some curious surprises when we look at the sort of obstacles that the wind passes over. Solid obstacles, like walls, do not stop the wind as effectively as partial obstacles like some fences and bushes. If we think of an obstacle in terms of how much daylight that we can see through it, then this gives us a rough measure of an obstacle's density. A brick wall has a density of 100 percent, as no light will get through, but a row of thick bushes might have a density of 50 percent. Bizarrely, the wind is dampened more effectively by obstacles with this middle "bush" density than by the solid wall. So when we look at the water downwind of some leafy trees we may see a calm area in the wind shadow that is actually bigger than it is down-wind of a building of similar height nearby. Another curiosity is that the wind will usually be lightest at a spot that is about five times the height of the obstacle downwind, not where we might expect it, immediately downwind of the obstacle. In most situations the wind will be back to about three-quarters of its strength at a distance ten times the height of the obstacle.

We don't need to memorize distances, heights, or factors; we can just enjoy looking at the ripples on the water and noticing how they change the further downwind we look from trees and other barriers.

The next thing to look for is ripples going the wrong way. It is time to think back to the eddies in the river, the way water spins as it passes any hindrances and can end up flowing in a thin line at the edge of the river in the opposite direction. Lakes are inevitably in bowls of land and this means that as a breeze passes over the top of the land around a lake, the wind will often form a circular vertical eddy just downwind of the high land. This in turn can create very local breezes at the level of the lake that are blowing in the opposite direction to the main wind. It is surprisingly common to find yourself standing by a lake feeling a breeze and watching its ripples blowing across the lake in one direction and the clouds above heading in the opposite direction.

Eddies can form, in a much smaller way, downwind and even upwind of obstacles. The air will rotate vertically, or roll, if it is going over the top of something, or swirl horizontally if being forced around a corner—think of those dust devils that appear near corners of buildings. (The main reason that trees often dampen the wind more effectively than walls is that they create fewer eddies.)

Finally, it is worth knowing that there will be some local breezes, as the sun warms the land and the water differently. We will go into these more thoroughly when we look out to sea, but for now it is worth knowing that if your lake is in a steep valley and it is a warm sunny day, then there will be some local winds created on the sunny side of the lake that will not be found on the shaded side.

If we bring these effects together, it gets much easier to see why we often get such intricate patterns of ripples and calm spots on the water. The wind will always obey the laws of physics, so the puzzle of what causes these effects can usually be solved. If you give some thought to the direction the wind is coming from, plus the height and type of obstacles it has to pass over, you will be able to have a lot of fun deciphering the patterns you see in the water. Just by looking for them you will notice a lot that passes others by.

At the water's edge, near a group of willows, I found a cloud of insects flying in an excited frenzy. There were the tiniest of dimples and then expanding rings as they dipped into the water. It is not just the fish keeping an eye on these tiny vibrations and the ripples at this micro level. Some insects, like the familiar Backswimmer with four legs touching the skin of the water (its other two legs are longer and are used as paddles), will be perfectly alert to the vibrations that signal another small insect in trouble. So one miniscule impression in the water can trigger a chain reaction in the life around it and lead to a flurry of commotion on the smallest level. This is very satisfying to watch; the moment that the calm stasis is broken, a brief frenzy ensues, and then calm returns for the cycle to begin again.

You can trigger a different kind of flurry sometimes. These surface insects are so sensitive to the slightest vibrations that a stamped foot by a patch of perfectly calm water will often trigger a collection of ripples emanating as insects dive or take off. Disturbing the insects' peace in this way is a guilty pleasure of mine.

I WALKED AWAY from the lake and as I did I glanced upward, hoping that one of the bigger, lower clouds might pass over the water. There is a technique that the Pacific Island navigators use out at sea, when searching for land, that we can enjoy using on land, when near large lakes.

As the Pacific navigator searches the sea for signs that an island is close by, they will also occasionally look to the sky for help. Land heats up more quickly than water, which means that clouds form over islands more dramatically than over the surrounding water. These cloud signposts can be seen from many miles away and are very useful. If the rising air currents are strong enough they sometimes split the cloud over land in two, creating what has been nicknamed by navigator David Lewis and others as "eyebrow clouds."

In theory we will get fewer clouds over lakes than we do over the land around them, but you are only likely to notice this by the largest lakes. The Islanders also study the undersides of distant clouds carefully, hopeful of spotting a color change. Clouds over surf and coral sand appear unusually light and white, over lagoons they are colored green, over dry reef they have a pink tinge, and over forested land they are darker.

When David Lewis was sailing with Iotiebata near the islands of Maiana and Tarawa, he could clearly see the green hue reflecting from under these clouds. Lewis could see the effect so distinctly that he wondered why Iotiebata had failed to notice it. Iotiebata was embarrassed to be asked and replied that he had not mentioned it because he did not want to patronize Lewis—it was, he said, so glaring a clue that "even Europeans can see this very obvious sign."

We can have some fun by turning this method on its head. By studying the undersides of clouds we will occasionally spot a subtle color change as they pass over lakes.

Lakes offer some of the best opportunities for us to notice how our awareness of water is growing and they resonate very strongly for me personally in this way. As a young man I would charge past large bodies of water in places like England's lake-filled region, not entirely oblivious to the beauty of the landscapes around me, but certainly blind to the intricacy of that beauty. I did not see how deciphering ripples could enhance my journey to a local peak.

It would take thousands of paces to reach the places I planned to get to and back. One of those steps would have triggered a reaction in a swan, which in turn would create a unique pattern in the water as it sent ripples to interfere with those of a wind eddy. Had I the awareness to notice this in the first place, it would have been viewed as a delay at best. I may have delighted in the freshness of the air and scents on the breeze, but I would never have spotted that these scents fluctuated in a way that was etched into the water beside me.

Now I view lakes and their patterns in a different way: They are a summit in their own right. If I take the time to notice the behavior of the birds, fish, insects, and breezes through the map of their actions in the water's surface, this becomes a conquest of another kind. A higher summit.

# 8

# The Color of Water

IMAGINE YOU ARE SITTING on a boat out at sea and a friend asks you the following question:

"What color is the sea?"

You glance around you, just checking that it is as stupid a question as it sounds, before confidently replying, "Blue. No, wait a minute . . . Green . . . Or maybe gray."

At this point your friend leans down, dips a glass into the sea, pulls it out, and puts it in front of you. You stare at this glass of perfectly clear liquid and briefly consider your choice in friends. Then your mind turns to the color of the water at your favorite rivers, lakes, and coastlines, and you realize they are all slightly different.

The variety in the colors we see when we look at water is one of the reasons we love it, but far more people enjoy this array than give any thought to the reasons for it. The Celts certainly appreciated the challenge of trying to describe the color of water. They cheated by using a prefix, "glasto-," to denote things that were blue/green/gray.

There are several parts to understanding water color. Each of them is very simple when considered on its own, but together they gang up to confuse people and make the subject appear much more complex than it needs to be. The four areas we will consider are: what is beneath the water, what is in the water, what is on the water, and the effect of light. The last of these, the relationship between light and water, we will begin to look at in this chapter and then, because it is a deep, rich, and fascinating area, we will explore it further in the next chapter.

THE FIRST THING to consider when trying to understand the colors you see in water is to work out whether you are looking at the water at all or whether you are actually looking at a reflection; sometimes this is obvious, but not always. Stand over a puddle and peer into it from directly above, and if the water is clear, you will be able to see several things. You might see your own reflection, you might see the ground below the puddle, or you might see some brown mud particles swirling in the water, particularly if someone has just walked through it. Each of these views is important in its own way, but the first thing to be aware of is that when we peer down into water, we have a choice of what to focus on, and this is because we are looking at it from a high angle, near the vertical.

However, if you now take twenty steps away from the puddle and look back at it, you will not see any of the mud or the ground beneath it; in fact, you will not see into any of the water itself, because all you will see are the reflections of whatever is at the same angle that you are looking from, on the other side. We cannot see into water at all if we view it from a low, glancing angle. This is very important when we're thinking about the color of water, because in many circumstances when we think we are looking at the water, we are actually looking at something different and in the distance. Looking out to the sea in the distance is a great example: What we see in that situation is dominated by the reflection of the sky even further in the distance. This is why the distant sea appears blue in fine weather and gray on overcast days.

Staring down into a puddle or gazing out to sea are two extremes, and it is easy to predict what we will see in those situations. But it gets a tiny bit more challenging when we look across water that is both close to us and a little further away. When we stand on the bank of a broad river we will be able to see at least part of the way down into the water nearest our feet, but we will only see reflections from the water on the far side. This is why the colors will often appear dramatically different from one bank to the other. You will notice this if you look for it, but not if you don't, because our brain has gotten used to this effect and so doesn't register it as at all peculiar.

Try scanning your eyes slowly from one riverbank to the other to see if you can find the area where the shift takes place from looking at only reflections—the far bank—to being able to see into the water (it is easiest to gauge this effect if the river's water is fairly clear). This transition will not take place in one exact spot as there are a few factors affecting it, including the angle and amount of the light that is coming from above and beyond. You should be able to notice that there is a patch where the shift takes place quite dramatically, (it is usually between 20 and 30 degrees or two or three extended fist-widths below horizontal).

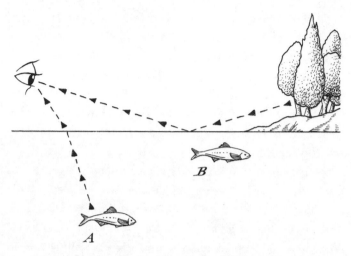

*We can see fish A, but see trees instead of fish B.*

THE SIMPLEST EXPERIMENT imaginable will prove that there is a fundamental relationship between light and the color of the water we see. At nighttime, place a glass of water in front of you, then turn off all the lights and the glass of water will disappear along with everything else in the room. The color of the water is black! That sounds so ridiculously obvious that it's almost pointless, but there is a point and it's this: Without the light hitting the water, it has no color at all. Light gives water its color.

Things become much more intriguing with the lights back on, when you compare the color of the water in a glass with the color of water in a bath. If your bathtub is plain white, try filling it until it is only a couple of inches deep and then peer in. You will see that it appears as totally colorless and transparent as the glass of water. Now continue to fill the bath until it's as deep as sensibly possible. If you look at this slightly deeper water, can you see the faintest of blue tints has crept into it? That blue tint in the bath's water is the reason that we see a blue sea when we look down from a boat into deep clear water.

Pure water is not colored, but it does absorb some color. When white light hits water, some gets reflected and some gets absorbed by the water molecules. The white light that enters the water is made up of all the colors of the rainbow, and the colors do not get absorbed equally. The reds, oranges, and yellows get absorbed by water more than the blue colors. The result of this is that the more water that white light has to travel through, the bluer it will appear when it emerges again. Have you ever noticed how a white object dropped in what appears to be clear, deep water turns a blue color before disappearing? If you pull a white bath plug from the bottom of a deep bath the opposite happens—it goes from the faintest blue to white.

As the scale increases from bathtub to swimming pool, the amount of water the light has to travel through grows, and so more red-to-yellow colors get absorbed, which is why we think of swimming pools as light blue, even if the bottom is white. Scientists have worked out the exact color that can travel furthest through the water without being absorbed: It is a blue-green color (They have even worked out its wavelength: 480 nanometers).

THE NEXT AREA to consider is what lies under the water you are looking at. The bathtub experiment works best if you have a plain, white bath for the simple reason that in shallow water, whatever is beneath the water will have a huge impact on the colors we see when we look at the water above it. On the beach you have probably noticed how the water gets darker from the spot where the waves break out to the deeper sea in the distance. But we are also used to the way each beach's waters look unique, with its own trademark array of colors. This is because the seabed will color the water, and the shallower the water, the more impact it will have.

If you are on a beach of white sand, the shallowest water, where it barely covers your feet, will appear white, but not far away there will be water that is very light blue, farther still it is a slightly darker shade of blue and as your eyes move farther from the shore, the color darkens until the lightening from the sand is not a factor at all.

On beaches with golden sand or pebbles you will see the same effect, but the colors will be closer to blue-green, or turquoise, where the yellow and blue mix, and it will shift from the lightest to the darkest shades with depth. In terms of the seabed it is a simple case of mixing of colors, like on a palette; the deeper the water, the more blue that is added, but in shallower water the more the seabed color is mixed in.

Now we're ready to combine three effects. If you stand in the sea, ankle-deep, you will be able to mix the color palette for yourself. Experiment with looking straight down, a few feet in front of you, and farther out, and try to sift the colors in relation to the blue tint of the water, the hue from the bottom, and the reflections from the sky.

On a sunny day with a scattering of fair weather cumulus scudding across the sky—the fluffy sheep clouds—you'll get a great chance to witness the effect that the sky is having on the color of the water you see. Notice how much closer to blue the water is in full sunlight compared with the water under a cloud's shadow. The effect is so dramatic that many people confuse cloud shadows on water with something big happening underwater; it is very common for people to think there must be a sudden change in depth or even a huge shoal of fish, but with patience you can watch these darker, less blue patches march in tandem with their parent cloud.

If you're not sure, try drawing a line from the sun through a small cloud down to the water. It is tricky to estimate exactly how far from you the shadow will fall, but this method will get you looking in the right direction at least. Equally, if you spot one of these darker patches you can trace a line up toward the sun, and if it's a cloud's shadow, there will have to be a cloud somewhere on that line. It is strangely satisfying to recognize the shape of the shadow in the water in the shape of its parent cloud.

Learning to recognize whether a color change in the water is a clue to a change underwater or in the sky is an art you can keep refining for many years. And in many parts of the world, like the Pacific, it becomes a very practical art. Where boats move between coral atolls, the best chart a boat can have is in the color of the water all around. GPS together with the latest electronic charts still fail to beat a knowledgeable local skipper whose eyes are tuned to the faintest lightening of the blue water. Even an electronic depth gauge, like an echo-sounder, is pathetic compared to these color changes, since it can typically only measure depth beneath the boat, not tell you how depth changes all around you. And electronic charts of coral atolls are still woefully vague and inaccurate compared with color changes.

After years of practice, navigators in these areas can pick a channel out through the treacherous reefs, just from a thin line of darker blue through the various hues of turquoise. A fisherman interviewed in the 1880s was asked about navigating in the North Sea and remarked:

> There's nothin' in the world can be easier, when you've learned your lesson, than to pick your way about in the North Sea just with nothing else to guide yer than the depth o'water an' the natur' o' the bottom.

The difference between the color of the sea in deep water and shallow water has led to a common nautical expression, "blue-water sailing." If someone says they are a blue-water sailor, then this is not quite as romantic or fatuous as it sounds; it is a reference to sailing in deep water, typically trans-oceanic sailing. "Brown water" is not used quite so much now, but it refers to shallower water, the convention being that brown water is the sea up to 100 fathoms deep and blue water is anything over that.

If you do notice a change in color in the sea and it proves long-lasting and immobile, allowing you to rule out clouds or light changes, then this is a sign that there is a genuine difference in the ground beneath the sea. If it is not clear what this is, it invites a little detective work; local sailors are likely to be able to supply the answer, but you can also do some sleuthing by getting hold of the local charts for the area, either in paper format or online. If it is your local or favorite area of coastline, it is a worthwhile investment to buy the local chart, as it will tell you so much and explain so many subtle shifts in color that might otherwise tease for years. A few scattered letters on the chart, "S" for "sand," "Sh" for "shells," "M" for "mud," "Wd" for "weed" and the rather wonderful "Oz" for "ooze," will reveal the nature of the seabed—critical knowledge for any vessel wanting to drop anchor—and this can help crack many a sea-color mystery.

ON A RECENT VISIT to the Isle of Purbeck, along the English Channel in Dorset, I spent time walking slowly around a place called the Blue Pool. On the sunny morning I was there, the water ranged from a vibrant turquoise to a more conventional, dark blue-green and, in the trees' shade, a much darker lake-green color. The pool has this name because the color of the water stands out. It is different from all the other lakes and the sea nearby. But since the bottoms of many of the other local lakes are similar, their depths are comparable, and they share the same sky, the reason for this special blue must be found by looking for some other explanation.

The Blue Pool is a disused clay pit, and it is the clay particles in suspension in the water that give it those alluring colors. Less alluringly, the Red River in Cornwall, in South West England, earned its name from the iron-rich tailings that were washed away in the heyday of tin mining in the region. This red color has left the water now that the mining industry has shut shop, which was good news for the wildlife, if not the miners, as it was too toxic for nature to thrive in the red water.

All the water you see outdoors will have some particles in it, even apparently pristine water sources, and the wildest of lakes will have millions of tiny specks both in and on the water. Algae, bacteria, dust, pollen, and other substances will be there and will be adding color to the water, sometimes imperceptibly, occasionally dramatically.

I remember looking out the window as the aircraft banked shortly after taking off out of Heathrow recently. Beneath the wing there were lakes that were too symmetrical, with too many straight lines, to have been entirely natural. I was struck by the way one of the lakes was a bright-green color and the others were not. The bright-green lake was nearest the neighboring farm, which explained the color anomaly.

Whenever the still water of lakes or ponds suffers from what scientists call "eutrophication," which means excessive nutrient enrichment, the delicate ecosystem gets tipped over. Algae needs three things to thrive: water, sunlight, and nutrients, and if it gets all three in the levels it likes then it will bloom, completely changing the color of the water.

Algal blooms don't just change the color of a lake or pond; they impact its whole ecosystem, drastically reducing the light that penetrates the water and using up a lot of oxygen that is dissolved in the water, potentially killing fish and other organisms. These bright-green waters are interesting to look at, but they are not the healthiest of habitats for plants or animals. Most people find them uninviting to swim in as they instinctively feel them to be impure and unhealthy, and indeed they can be toxic.

In some parts of the world, a more flamboyant display is put on by the algae. The Amazon is a yellow color in places and near Manaus in Brazil this yellow water meets and mixes with the reddish-black water of the Rio Negro, creating a vibrant scene. Each color is the result of the rivers having different particles and algae in their water. One particular alga, *Dunaliella salina*, turns some salty lakes, like Lake Hillier in Australia or Lake Retba in Senegal, an unbelievably bright pink. "Red tide" is the name given to a number of algal blooms that change the color of the sea to a red or red-brown color. They are believed to be caused by fluctuations in nutrient levels and temperature, and they are potentially toxic to marine life and humans.

At the opposite end of the scale, if the water is "oligotrophic," meaning very low in nutrients, it will be especially clear and inviting as the algae cannot get going. One of the biggest examples of this is the Mediterranean, which is especially low in nutrients, leading to very low levels of algae and the crystal clear waters so many love to holiday by and swim in.

A lot of the particles that we see in water will be inorganic, a mixture of mud, sand, clay, silt, chalk, and other substances, each one affecting the colors we see. This is typical in lowland rivers, where the motion of the water is forever churning up mud and silt from the banks and making the water an opaque, light-brown color. Whenever you see a tributary joining a stream or river of a different size, compare their colors and you will spot that they are rarely identical, as their journeys and therefore the particles they have picked up will be different. In Borneo I spent time with the Dayak and witnessed their ability to read the very finest details in the rivers—they recognize color changes at stream junctions as clearly as we read road signs.

Some glacier-fed lakes take on a bright, light pastel-blue color because of something in the water known as "rock flour"—small rock particles that have been finely ground by the pestle and mortar action of the glacier and then fed into the lake. These unusual waters can be found wherever there are glaciers, including Canada and Norway, and are healthy despite their vibrant appearance.

If you get the chance, study the sea close to the shore before and after a storm. The violent upheaval often whisks the seabed up and imbues the water with a new color, which will fade over the course of a day.

THE COLORS WE SEE will sometimes also be affected by what lies on the surface of the water. The iridescence of an oil spill makes for alarming news bulletins, but we can spot the same optical effects from the organic oils on the surface of a cup of tea (it's easier in tea without milk, and while facing toward the light, with your eyes low). Oil, dust, and many other temporary visitors to the surface of the water change the water in a way that is generally seen as unwelcome, but there is one frequent guest on the water's surface that is considered beautiful.

Both the Greek and Roman goddesses of love, Aphrodite and Venus, sprang from the foam of the sea; the mythic moment was captured by Botticelli in his painting *The Birth of Venus*. Like paparazzi door-stepping the most private moments of a classical goddess, Botticelli gives us a glimpse of things that are only meant to be imagined. But what Botticelli failed to include in his painting was that the foam was, according to Hesiod, the result of Uranus's genitals having been cut off by Cronus and thrown into the sea. Stepping away from the slightly sordid mythology, the science behind foam is rewarding.

Foam will add color to the faster stretches of rivers and seas under strong winds or breaking waves, which makes it appear everyday, and in a sense it is, if you are on Earth. Scientists believe that Earth may be the only place in the solar system where water blows over open water to create foamy waves.

We know the foam we see is usually white, even though the water isn't. In fact the foam is still white on water that is a dark, muddy brown. Even the bubbles in Coca-Cola are white, but why? For the same reason that clouds are white and most powders are, too. Have you ever noticed that if you grind something colored into a fine powder, it often turns white, regardless of the color it was to start with?

Foam is made of tiny pockets of air surrounded by water. Inversely, clouds are made of very small water droplets surrounded by air. When light hits either of these it meets a collection of spheres of varying sizes. The light gets bounced off these different sized "balls," and each size reflects a different color. These colors arrive at our eyes simultaneously, and what we see is them mixing together again to create white light. The same thing happens once powders are fine enough. But look closely enough at foam and the colors will sometimes reappear fleetingly.

The bubbles in foam will typically burst very quickly and the foam will soon disappear, something we will all have witnessed around our feet on the beach. Foam that lasts longer is a sign that there is something else in the water, specifically chemicals called "surfactants." These compounds, which include common soap and many comparable industrial chemicals, make the bubbles last much longer. Long-lasting foam is a certain sign that the water is not pure.

Romantics look away now!

In an earlier book, *The Natural Explorer*, I wrote about an instrument that was invented by the Swiss traveler and physicist Horace-Bénédict de Saussure called a "cyanometer." It was effectively a collection of colored swatches that could be compared with the sky color and then used to measure and label the sky's blueness.

If the idea that the blueness of the sky can be measured and given a number fills you with the dread notion that the empiricists will stop at nothing, then you really must stop reading here. Water scientists have developed their own version of the cyanometer, called the Forel-Ule Scale, as it was developed by the Swiss scientist Francois Alphonse Forel and the German geographer Willi Ule. The scale works like this: Twenty-one small test tubes of liquid are created and numbered, and they cover a scale of colors from bright blue (no. 1) to dark cola brown (no. 21).

As we have seen, the reflection of light off the water will have a massive impact on the colors we see, so the Forel-Ule method gets round this by lowering a white disc into the water until it disappears, noting that depth and then raising it to half that depth. The color of the disc is then compared to the vials of liquid and the closest number is assigned to that water.

Numerous tests and research have shown a couple of interesting things: Firstly and most surprisingly, we humans are actually quite good at objectively comparing colors, so we can give a reasonably accurate reading and therefore number. Secondly and most usefully, this color is a fair indication of what's going on in the water.

The following table gives you a rough idea of what we can deduce about the water from its color, once we've satisfied ourselves that this color is due to particles in the water, not the bottom or the light reflected off the surface.

- **Indigo blue to greenish blue** (1–5 FU scale). Low nutrient levels and low organic growth. The color is dominated by microscopic algae (phytoplankton).
- **Greenish blue to bluish green** (6–9 FU scale). The color is still dominated by algae, but also increased dissolved matter and some sediment may be present. Typical for areas toward the open sea.
- **Greenish** (10–13 FU scale). Often coastal waters which usually display increased nutrient and phytoplankton levels, but also contain minerals and dissolved organic material.
- **Greenish brown to brownish green** (14–17 FU scale). High nutrient and phytoplankton concentrations, but also increased sediment and dissolved organic matter. Typical for near-shore areas and tidal flats.
- **Brownish green to cola brown** (18–21 FU scale). Waters with an extremely high concentration of humic acids, which are typical for rivers and estuaries.

The above table is used by the Citizens' Observatory for Coast and Ocean Optical Monitoring, and for those who are intrigued and wish to get involved, there is a citizen science program and even an app. Horace-Bénédict, beat that!

# 9

# Light and Water

L AST YEAR I visited an exhibition of Constable's work at the Victoria and Albert Museum in London. "The Making of a Master" was a much-lauded temporary show that focused on Constable's techniques and development in the context of the grand masters. At times, to my shame, I found myself unmoved by the work on display.

Any feelings of indifference were swept away, however, when I stood opposite his 1796 painting *Moonlight Landscape with Hadleigh Church*. Looking at this picture I was struck by the extraordinary rendering of the water. The painting spoke to me in a personal, almost conspiratorial way—it felt as though I was receiving a secret code. For a couple of minutes, Constable and I were two members of a curious cabal. But then this strange society grew.

Hung next to this work was the one that had clearly influenced it: *Landscape by Moonlight*, painted more than 150 years earlier by Peter Paul Rubens. And in this picture there is another example of the phenomenon that had leapt out of

Constable's painting, different and yet also extraordinary in its faithful treatment of reflections on water.

It is rare for anyone to notice or care about, let alone faithfully reproduce the way the moon reflects off water, and I could not have been more excited to have found myself opposite two such brilliant examples. In each picture we can sense what the water is doing, once we have learned to study the anatomy of an optical effect known as a glitter path.

I should not have been surprised by Constable or Rubens. This is what the great artists manage. They flatter us, by observing better than others and then speaking to each of us as individuals and in a language that we worry we may be the only ones left caring for.

*Landscape by Moonlight,*
*by Peter Paul Rubens.*

WHEN LIGHT STRIKES water and then reaches our eyes it must have followed one of three paths. We only see the light because something has caused it to change direction dramatically and find our eyes; it will either have been reflected off the surface, off the bottom, or off particles in the water. In the previous chapter we looked at many of the effects that occur when light gets bounced around and reflected off particles, creating colors from postcard blues to worrying pinks. In this chapter we will delve a bit deeper into the light that gets reflected off the surface and the light that makes it all the way to the bottom.

Seek out a spot where you can get close to deep clear water and you can both look down into it and look across it toward a bright sky. If you can, find a place where the water is fully in shade, right next to a spot where there is no shade. This will demonstrate how important the light levels are, because looking at the shaded spot you will be able to peer down into the water, but peering at the bright area you will struggle to see below the surface.

Next find a spot where there is a mixture of shallow water in the shade and water that is lit brightly by the sun from behind your shoulder—early or late in the day are great for this. Can you see how things have swapped and the brightly lit areas are now easier to see underwater than the shaded areas? The bright light from behind you does not create glare on the water and instead is allowing you to see lots of detail underwater and on the bottom.

Once you've appreciated this stark difference between water in light and shade, you will have worked out how to spot fish, plants, and insects underwater that you would otherwise miss. There is a riverside spot I love on the banks of the river Arun, in Southern England, and one of the reasons I head there is that there are some mooring piles, tall pillars that rise out of the water near the riverbank and cast a series of shade stripes on the water. I love using these pillars and their shade to watch fish appear and disappear as they move from shade to bright to shade.

ON A SUNNY day, if you stared into a calm, empty swimming pool and focused your gaze on the bottom, you would be able to see it clearly. It would not be doing anything unusual or surprising. If you then waited until someone had just jumped into the pool and took another look, it would be hard to see the bottom at all clearly because of all the commotion on the water's surface. But once that person has got out and the water calmed, take one more look because you will see something beautiful. There will be wonderful patterns of light, bright white dancing loops on the bottom of the pool. Look at the underside of a bridge on a sunny day and you will often see the same patterns, flexing bright shapes projected up onto the dark underside from the water.

These two effects are related and demonstrate an effect in two of the areas we're interested in. The patterns of light on the bottom of the pool are formed by light that has made it through the water to the bottom, but the strange shapes that are constantly reforming are the result of the gentle waves on the surface of the water. These undulations make the surface of the water act like a flexing lens, focusing the light in some places and away from others, creating the pattern of very light lines and darker spots. The pattern under the bridge is caused by the sunlight being reflected off these lens waves in the river—you will see the same effect on the hulls of boats on sunny days, too.

When the sun is still high in the sky, its reflection is too bright to be safe to look at directly for any extended period of time, but if it's a moonlit night, you can look at the moon's reflections. Watch how the reflected light of the moon forms intricate patterns on the water, similar to the ones under the bridge or on the bottom of a swimming pool. These patterns are known as "moon circles" and can be seen when any bright point of light bounces off water. The way the shapes of these moon circles change is too quick for our brains to be able to track properly, but time-exposure photography has shown that they are formed when a tiny spot of light—a miniature moon—splits into two bright spots that dance away from each other and then glide back together and vanish.

The swimming pool effect is easy to spot and you will be able to find it in clear rivers, lakes, and sometimes the sea, too. And if you take the time to look for it you will find some surprisingly subtle incarnations out there. Anything that lands on the water will bend the surface skin of the water slightly, even the tiniest of insects. On a sunny day, an insect resting on a clear, shallow pond will create a beautiful pattern on the bottom of the pond: tiny, bright pools of sunshine, one for each of its tiny feet—this is one of my favorite freshwater sights.

So much of what we see in water will be influenced by the way the light and the surface of the water interact. The best way to prove this for yourself is to do a little observation experiment the next time you think you are looking at water that is calm: a placid pond, lake, or very calm river—even a swimming pool will do. The water may appear to be perfectly still and calm, but that is only because we are not standing in a place that allows us to spot the finer motions. We can always detect motion in the surface of the water if we find places where light and dark reflections mix.

All you have to do is explore near the edge of the water, perhaps taking a few steps back, until you find a spot where you can see the line where the reflections of both a light area, like the sky, and a dark area, perhaps trees or buildings opposite you, meet each other. At the edge of these two areas, light and dark, you will be able to clearly see how busy the water is: little swirls, baby currents, insects or fish creating the tiniest waves. The surface of water outdoors is never static; if it appears motionless then that is just a sign that we need to move to find the spot that blends light and dark reflections.

If you now move again until that spot you were looking at is either in an area of all dark or all light reflections, then it will appear to be relatively calm again. This experiment shows how important reflected light is to our ability to read what relatively calm water is doing. It is well worth spending a few moments seeking out this line between light and dark reflections whenever you are interested in what the surface of water is doing, as you will often learn as much in a minute as you would do in an hour of staring at either wholly bright or dark patches. It is very satisfying to be able to detect a fish below the surface from the slightest of fluctuations in water.

Something you will notice after looking at reflections in water for a while is that water does not act as a perfect mirror. Light objects will appear a little darker and duller in their reflected images and dark objects will appear a touch lighter. There is another important difference, one that surprises many, not least beginner artists: a reflected image gives a slightly different perspective of the object you are looking at. A reflection shows you the view from the point on the water's surface that you are looking at, not the perspective from where you are standing. It may sound complicated, but it's straightforward when you get out there and look for it. One way of thinking of it is that the reflection reveals more of the underside of anything that is in or very near the water, like a low bridge or the bottom of a duck standing in very shallow water. For this reason, I like to call this the "Duck's Bottom" effect.

*The Duck's Bottom effect.*

My favorite example, because it helps me practically at times, is the way that the trees we see on the opposite side of calm water and their reflections give us two different views of the same trees. In natural navigation it is very useful to be able to read the shape of tree branches (they grow more horizontally on the southern side and more vertically on the northern side), but sometimes it is hard to make out the shape of tree branches if there is no sky behind the trees, only more dark trees. In this situation the tree's reflection in the water can be very helpful, because it means you can look at your tree from lower down, which often holds them more up to the sky for you.

*Ripples make horizontal shapes disappear before vertical ones.*

If the wind picks up and there are any big waves, you will lose the ability to make out any detail in the reflections, but if it is breeze over a rippled surface there is an intriguing effect worth looking for. In rippled water, notice how you can often make out the shape of vertical structures reflected in the water, but not horizontal ones. The prime example of this is a bridge with pillars. If the water is rippled, you will usually be able to see the reflections of the bridge's pillars in the water, but the main bridge itself disappears almost completely.

One of the most beautiful effects of light on water is the way the sun forms a bright pillar as it is reflected off a large body of water toward the start or end of the day. This long line of shimmering reflections is known as a "glitter path" and is caused when our eye picks up thousands of tiny sun reflections of the sides of waves stretching into the distance. The shape of the glitter path is a measure of how high the sun is and the roughness of the waves. The glitter path will get narrower as the sun gets lower and broader as the waves get steeper.

Mathematicians have worked out that this is a fairly exact science and glitter paths reveal certain things about the sun and water with precision. For example, if the sun is 30 degrees above the horizon and the waves are angled 5 degrees, then the glitter path will be 20 degrees long and 10 degrees wide. But we don't need to immerse ourselves in the math to find glitter paths meaningful, we just need to watch them and keep the two basic rules in mind: The height of the sun and the roughness of the water will change the length and width of the glitter path.

In practice, what this means is that when you look out at a glitter path, the height of the sun isn't going to change noticeably over a few minutes, so any change in the width of the glitter path is a sign that the wave steepness must have changed. It is very common to spot that the path is not an even width and that it sometimes bulges noticeably in places. Since the glitter path will get wider when the waves get steeper, this is where there is a local rough patch of water, perhaps an area that is more exposed to the wind.

*The glitter path is narrowest where the water is calmest.*

THE SUN does not hold a monopoly on glitter paths and they will be formed by any bright source of light, if it is opposite you and low enough: the moon, planets, or even the brighter stars all form glitter paths at times. Some of the ones you will see most regularly are made with artificial light. I remember looking out across Falmouth Harbour in South West England one night, watching the boats come and go and enjoying noticing how the busy city's lights were forming long, thin glitter paths in the water. It was very rewarding to watch one bulge in the middle where the water was being ruffled by tidal currents.

It is quite common for the water to be rougher nearer where you are standing than farther out, as the water will be shallower nearer land and this will cause any waves to steepen. This often creates a broadening or fanning of the glitter path closest to you. In fact, this effect is so common that our brains have grown accustomed to it, so much so that I've noticed how computer graphics designers use it cleverly in computer games, to trick us into thinking the water looks more real. The irony of software designers tricking us into believing water is real by observing nature more carefully than we often do is quite amusing and perhaps a sign of the times, or things to come.

Glitter paths normally form a straight line between you and the sun or moon or other bright light source. But sometimes you will notice a slight bend or curve in them. This bend is not the same as a bulge due to steeper waves, which will normally be equal on both sides, i.e., symmetrical, but is a curving of the path itself. The most common cause of this is the wind blowing across the glitter path, changing the shape of the waves. It is a beautiful effect, well worth looking out for as it feels almost as though you can see the wind blowing sunlight across the water.

Glitter paths will usually appear redder in color than their "parent" light. A cream moon may give an orangey glitter path, as the light at the blue end of the spectrum is scattered away, leaving only yellows, oranges, and reds.

*A ripple map around a small sand island . . .*

*. . . and one forming around water lilies.*

*Noticing the plants that thrive in wet places, like these soft rushes . . .*

*. . . makes it easy to discover hidden water.*

*Duckweed, a sign of slow water and high levels of nutrients,
often from waterfowl or a nearby farm.*

*Eddies will form when water flows past any obstruction,
even when it is moving very slowly.*

*A series of miniature "pillows" and "holes."*

*Riffles and pools on a river.*

*Caddis fly larvae* [Agapetus fuscipes]. *A sign of very high-quality water that has not been polluted or run dry in the past year.*

*A shady pocket on the far side of the stream. A likely lurking place for a trout.*

*A "kissing" or "sipping" rise.*

*A "kidney-shaped" rise.*

*The colors we see in water depend on the brightness and angle of the light and the water's depth, as well as what's on, in, and under the water.*

*Dark acidic soil often holds clear oligotrophic water.*

*The angles we look from and the light reflections determine whether we can see underwater. Gently disturbed water acts as a lens, forming patterns on the lake bed.*

*Reflections appear a shade darker than original. There is also a subtle "Duck's Bottom" effect.*

If you don't see a glitter path when you would expect to, this is a sign that the waves are too big and steep. You will always see some reflections in water that is too calm for a full glitter path, like a wobbling stretched moon or sun, or even something close to a true resemblance if it is flat calm. But when waves grow beyond a certain point, the glitter path will disappear, as the faces of the waves no longer act as mirrors for us. If the waves don't look at all big, then maybe try taking off your sunglasses—glitter paths are highly polarized, so they appear greatly diminished when seen through polarized sunglasses, which of course is part of the reason sailors use them.

If the glitter path you are looking at stretches across the sea and an area close to land, maybe an island in the middle distance, look especially carefully. You should be able to pick up some of the areas where waves have hit the land and reflected, and that are causing the interesting interference patterns that we saw in the water in the Introduction. These will either narrow or broaden the glitter path, or in certain situations, like where waves meet perpendicularly, this will create slightly unusual patterns, like rectangular grids of bright, white dots.

The subtle beauty and intricate architecture of glitter paths is hard to recreate from memory, which is why it catches landscape artists out regularly, particularly those who depend upon their imagination a touch too heavily. Once you've spent a bit of time admiring glitter paths outdoors and in photographs, you can quickly spot an artist who hadn't seen the things they pretended to have seen. There are countless examples of contemporary artists who render glitter paths completely inaccurately— even more extraordinary in the age of photography.

You don't need the full glory of a glitter path to notice the way the wind will influence the light you see on water. Here's a useful trick: On fine cloths like felt you can tell which way they have been brushed by the shades you see—even a hand passing lightly over them will leave areas of light and dark. If you run your hand over a pool table, a suede coat, or any velvet, you'll be able to leave and see these marks. These areas of light and dark are not random; they follow a simple rule, which is that if the cloth gets brushed toward you it will appear darker, and if it gets brushed away from you it will appear lighter (this is the same effect that creates light and dark stripes on mowed lawns; the grass appears darker when the mower has been moving toward you).

When we look at water, it makes a huge difference whether the wind is blowing toward us or away. The easiest way to spot this is if you are looking out across a large body of water when the sun is not too low, and facing into the wind. What you will notice is that the water appears slightly darker directly in front of you than it does to the sides, because this is where the "knap" of the water is being "brushed" most directly toward you. The opposite is of course true when the wind is behind you. The water will appear a shade lighter directly in front and a bit darker to the sides.

There is so much detail out there that our brain has evolved a few shortcuts, and while these are very useful, even essential, most of the time, they can create one or two surprises. People have seen human faces in the craters of the moon, the face of Jesus in toast, and a monkey in the bark of a tree. In December 2014 when a storm hit the UK (a storm that the media found too exciting to call a storm and instead labeled a "weather bomb"), observers found a face in a photo of one of the waves that was described variously as "an elderly gentleman," Scrooge, or God.

This habit of our brain to find patterns and ascribe meaning where there may be none is called "pareidolia." It is a fascinating and at times bewitching habit and one that all water readers should bear in mind before racing around and proclaiming the miracle. It does not mean the face of Mary Magdelene that you saw in the reflection of the streetlights in a puddle isn't real, but we just need to consider the possibility that it might not be.

Sometimes light and the particles in the water play together to create intriguing colors and unusual effects. You will never see your own shadow in pure water, because there is nothing for the light to bounce back off. You might see your shadow on the bottom of a pond, pool, or stream, but you cannot see shadows in pure water itself. However, when water has particles in it, it becomes cloudy, or "turbid" and then light can bounce off these particles, which makes it possible for us to see both the light shining into the water and shadows, too. If it helps, think of a flashlight beam in a dark room; you can see the wall the beam lands on, but you cannot see the beam itself, unless there is a lot of dust in the air, and then the beam is easy to see. So, if you see your shadow in the water, it is proof that it is not pure and has lots of particles in it.

If you are gazing down into cloudy water looking at your own shadow, there are a couple of extra effects worth keeping an eye out for. The first is that your shadow may have an orange-hued fringe around it. This happens because the tiny particles in the water don't reflect all wavelengths (and therefore all colors) back equally to you. Orange makes it back more easily than the others. The second effect, which, if you see the orange "halo" effect, is definitely worth looking for, is that you may spot shafts of sunlight emerging from your shadow and radiating out away from it underwater. This effect is sometimes nicknamed the "aureole effect." These radiating rays are not caused by any holy behavior on the viewer's part, but are the optical effect of looking in the opposite direction to the sun, which is what we are always doing when we look at our own shadow. In still water, your chances of seeing the aureole effect will be increased slightly if you stir the water with a stick. In deeper water, I have found it easy to spot in the Mediterranean.

I REGULARLY GET asked questions with a similar theme: Why, in the modern world, should we take the time to notice these things? The questions range across the spectrum of politeness, from the necessity of these skills to the pointlessness of them. Inevitably this makes me ask myself the same questions, often when peering into water. The answer I give myself and others is rarely perfect, but that is OK. The only reward is the excitement of seeing the world, for a moment, with the eyes of an ocean navigator's and Old Master's love-child.

# 10

# The Sound of Water

IN DERBYSHIRE'S PEAK DISTRICT, in central England, there is a village called Eyam with a wet and macabre history. The name "Eyam"—pronounced "Eem"— derives from the Old English word for an island, and the village is tucked neatly between two brooks. These days Eyam is much better known, where it is known at all, for being the village that sealed out the world as it suffered its own private hell, when the plague swept England in the seventeenth century. Water played a part in Eyam's tragedy.

When the village tailor, George Viccars, ordered some cloth from London in 1665, it arrived after its long journey a little bit damp. Viccars laid the cloth out to dry and the plague-carrying fleas, which had made their home in the cloth and hitched a ride from London, escaped. He died shortly afterward. As plague gripped the village, its inhabitants were urged on by their vicar, the Reverend William Mompesson, to do something selfless and terrifying: They went into lockdown. Nobody was permitted to enter or leave the village while the epidemic raged.

By October of the following year, 259 out of about 350 villagers had died. Controversially, William Mompesson had sent his children away from Eyam just before the quarantine was established and had wanted his wife to go with them, but she refused, wishing to stick by him. She survived until near the end of the epidemic, but then succumbed and died. The vicar himself was one of the minority to survive the outbreak.

Whenever water flows it creates sound, and that sound can be used to make a map of our surroundings. A village named for being surrounded by water seemed an ideal location to experiment with the possibilities of water-sound maps, and I headed to Eyam with a couple of specific places near the village in mind.

The villagers would have starved without some means of receiving help from the outside, and so there were a few designated places where food or medical supplies could be left and then collected. For payment, the villagers left sterilized coins in water, the metal first having been purged of plague with vinegar. One of these places of desperate exchange has become known as Mompesson's Well, named after the vicar who urged the village into isolation.

It may be called a well, but it is really a spring, a bubbling spring covered with stone slabs to create something resembling a drinking font in the middle of an otherwise featureless field. (To the untrained eye the field may have appeared featureless, but to the seasoned water reader there are always signs, because water changes landscapes. The spring feeds a weak stream that hugs the bottom of the gentlest of valleys in the field and its course was clearly painted by the lush dark-green of soft rush.) Sitting at the top of the field enjoying a sandwich, I watched how this low, wet land was a magnet and regular focus of the birds in the area. Behind me crows and magpies fought for space and bluebirds flitted to higher branches, their alarm calls signaling mild consternation at the squabbling corvids.

I walked away from the most babbling part of the spring, stepping slowly up the grassy hill and listening attentively after each step to the sound of the water. Water does not emit a single sound, of course, but a range of volumes and notes, most short but with an occasional longer one. I walked until the sounds disappeared and noted how this does not happen at one exact moment.

The sounds we hear outdoors are affected by the wind, the way the sounds spread out and are absorbed by the air, the shape of the ground, any obstacles, and the air pressure, temperature, and humidity. So the first thing to note about sound maps, not least water-sound maps, is that the information in our new map is different from every other map we are used to.

Certain factors allow sounds to travel farther, such as flat ground, few obstructions, and cold air. The closer to the ground we are, the more dramatic become the small changes in the shape of the land. Having reached the far edge of the range in which I could still hear the bubbling spring, I lowered my head little more than a foot and the spring disappeared off my audible map. This is an interesting phenomenon, especially on a slightly larger scale.

Something I have discovered by studying my own reactions and those of unwitting groups of walkers, is that the moment a distinct sound disappears is often also the moment that we begin to feel lost for the first time. It is an experiment I like to repeat around waterfalls, by asking people to point to nearby landmarks. While the waterfall is clearly audible, few fail in this simple task, but the moment the land gradient changes and obscures the sound of the water, their ability to orient themselves, not just with reference to the waterfall, but to all other local landmarks as well, often deserts them. This is especially true when someone is relying on the sound of the waterfall subconsciously.

In the diagram below, at point A everyone is usually able to point to local landmarks, even when they cannot see them, using the clear orientation they get from the sound of the waterfall. But at point B, the shape of the land has obscured the sound of the waterfall and a large proportion of people struggle with the very same exercise.

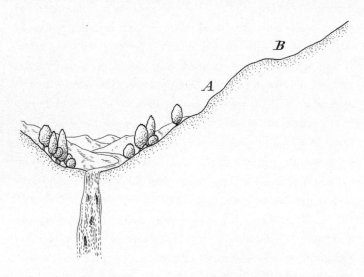

Low-frequency sounds, i.e., deep ones, are better at traveling around and through obstacles, but the higher-frequency ones tend to get reflected by things in their way. This is one of the reasons why it is the neighbor's "woofer" bass that drives us up the wall, not their violin. It is also why some police forces are experimenting with lower-frequency sirens: The higher-frequency ones can be confusing in built-up areas, as the sounds bounce off the buildings and appear to come from all directions.

Near the top of the small hill above Mompesson's Well, there were a couple of cars parked and occasionally I could hear the sound of the spring behind me reflected back. But it only happened when the breeze was just right and the stream I heard was a different character from the one I had heard at the bottom of the field, as only the higher notes were surviving the journey all the way to the cars and then back into my ears.

LATER THAT DAY, my walk took me to a place where I knew that there was a stream running down a gulley at varying distances from the path that I would be climbing. I paused at the base of the steep hill and listened: No water was audible, just the intermittent sounds of traffic behind me. A few steps on and the cars faded and were replaced by the sounds of hungry chickens in a yard. I walked up the hill for a short while before pausing again. Now a new sound reached my ears, children playing in the village school, their energy-spilling outdoor voices carried very clearly to me, even though I was now farther away than I had been a minute earlier.

I noted how their voices disappeared as the wind lulled and I used the satellite dishes on the houses, and the streaming chimney-smoke below, to work out that the invisible but very audible school must have been west-southwest of me. (Most TV satellite dishes point close to southeast in the UK, because the dominant broadcaster's satellite is in that direction.) But still I could not make out the sound of any water.

The gradient of the path softened and then flattened at a small plateau and the mud squelched under my boots in the spot where so many others had also chosen to rest for a second. Suddenly there it was, the unmistakable hiss and gurgle of white water. Like the voices and the spring earlier, at the edge of its audible range the water appeared and disappeared with every change in wind strength and direction.

This was a much stronger beacon than the small spring of Mompesson's Well, and I was able to use it to do more extensive experiments. The sounds that reached me changed every few paces, as they passed through a varying selection of trees and foliage between me and the water. There was a dappled range of pines, spruces, and beech trees between the path and the water, and each offered up a different sound filter. A clump of spruces had hundreds of horizontal lower branches jutting out, all needleless and interlocking; they reminded me of the teeth of a traditional music box. The sound of water through these was clearer and more even than the broken, muffled sound that struggled through the brown leaves of the beeches. Beech trees, like some oaks, willows, and a few others, hold onto their lower leaves in winter. This peculiar phenomenon is called "marcescence."

In any area of beech trees, look around in winter and you'll quickly spot how the higher branches have lost all their leaves, but the lower branches, those within a few yards of the ground, hold on to their brown leaves. Hedges still thick with brown leaves in the depths of winter are beech hedges; beeches are favored by hedge-layers for this very reason of not being denuded by autumn.

Scientists have done studies to work out the crops that scatter sound well, making them good sound barriers, and those that do not. It turns out that the long ribbon-like leaves of corn and wheat scatter sound surprisingly effectively. The sounds of water build a map not just through the volume, or the high and low frequencies we hear, but the timbre is a key to what is around us, too.

With each step I built a better understanding of how the sound of the water changed with each tree-type within the forest to my left. It is a technique that I have played with before in wilder climes, deep in the heart of Borneo. There, with little else to entertain me, wedged into the bottom of thin, wooden canoes for days at a time, I learned to read the character of the riverbanks with my eyes shut, by listening to the changes in echoes from the noisy outboard engine. Echoes from the typical mix of tangled roots and mud were like a distant shaking of tinfoil. When this echo changed to a harder, more percussive sound, like a power saw, then I knew we were passing a limestone bank.

I ENJOYED NOTICING how the sound of the falling water changed with every bump and wrinkle in the land. Using only my ears, I was ecstatic to discover an uprooted tree stump, complete with giant earthy root-ball, as the tumbling water was silenced for a second.

My mapping of the land in this way, by listening for water and unexpected quietness, has been shaped and more finely tuned by an example of Dutch landscaping ingenuity. Schiphol is the fourth-busiest airport in Europe and a very noisy place indeed. One day the local residents noticed something both peculiar and welcome: it became a lot quieter after farmers plow the surrounding fields. The landscape artist Paul de Kort was given the task of following the example of the farmers' fields to reduce the aircraft noise around Schiphol, and this led to a beautiful and original piece of landscape engineering around the airport.

It turned out that it was the ridges and furrows in the fields, and the angles that these created, that proved so effective at reducing noise for the neighbors of the airport. These ruffles in the landscape were bouncing the sound up into the sky and deflecting it away from those living nearby. And so de Kort set about creating a park of enlarged ridges, sculpted by GPS-guided excavators, to dampen the noise on a grand scale.

I walked over to the spruce stump, sat on it, and poured a hot tea from my flask. The tea was still too hot to drink—thermoses never cease to amaze me—so I continued to listen as I waited for it to cool. A twig snap drew my eyes down the wooded slope and an ivy leaf, severed by a squirrel's leap, rocked slowly down to the ground. I could hear the raging water, but it was still just out of sight.

Something that only occurred to me as I sat there, sipping my tea, is that we will hear a different part of any river with each small shift in wind direction. The closer the wind was to westerly, the harsher and raspier the water sounded, but whenever the wind backed to the south, the softer the water sounds became. The water must have been enduring more severe upheavals on the steeper ground to my west than it was to the south. Focused, attentive times like this remind me of the "still-hunting" method used by Native Americans, who would wait very patiently for their prey to stumble on them, rather than the other way round.

A few seconds after setting off again, I was shocked to find a small stream gurgling away just in front of me. This would not be a shock on any ordinary walk, but on a walk whose main purpose was to actively listen for, and to, water? It turned out that the hill had set up a perfect demonstration for me. I was walking with the wind on my back, over an undulating series of badger setts and sprawling spruce-roots with loud sounds of water behind me. Just as we cannot see what is behind us, we cannot hear what is shielded from us by wind or land shape.

The curious thing about this situation is that we normally think of things as being invisible when they are behind us, but when searching for water with our ears, the concept of "behind" is always relative to wind direction and topography, not the direction we are facing. Something can be right in front of us, as that small stream was for me, but not appear on our sound map if the wind is coming from behind us and the ground is uneven.

On a short leisurely walk, this may only have the effect of bending the things we sense slightly in favor of us noticing sounds upwind of us, but on occasion it can create a more dramatic warping of the map. During the American Civil War battle of Iuka in 1862, a northerly wind combined with the shape of the land to create an acoustic shadow. It turned out that two divisions of Union soldiers were positioned in that shadow and missed the entire battle, despite the guns raging only a few miles downwind of them.

My route was sinuous and circuitous. A long length of gray road led me back down the hill into the village where its water heritage was all too apparent in the street names: Water Lane, the Causeway, and Mill Lane. I noticed that the sundial on the church had markings not only for reading time, but latitude and season, too—a natural navigator's church for sure. Then I headed west out of the village, my ears now tuned and ready for the exercise ahead. They would be needed in my search for a mysterious waterfall.

A waterfall is the result of water dropping vertically from one level to another, often from a hard rock to a more easily eroded soft rock—so obvious and so simple. Yet, there are a myriad of different types of waterfall, named for the way they fall or are formed. Ribbons are taller than they are wide, punch-bowls start thin and blast out into a wider pool, fan waterfalls spread out, horsetails stick to the rock, giving a flowing white effect, segmented ones split on the way down, and both tiered and cascade waterfalls take it in stages.

I knew that near the village of Eyam there is a beautiful waterfall called Waterfall Swallet. A swallet is an archaic name of a depression or sinkhole in the ground, and it is fitting for this waterfall because the water flows as a stream at ground level but then disappears over a rocky ledge into a great hollow in the ground. We are used to water falling from higher up down to a normal ground level, but this waterfall ran from a normal level down an abyss. There had been heavy rain in the days before, so I hoped I could count on a gentle thunder as the water smashed onto its new, lower level.

This would be helpful for two reasons: The unusual landform around this particular waterfall meant that it is invisible until you are very close to it and, secondly, its location is something of a secret. There must be a fear that this little beauty spot will become swamped with tourists and trampled with human hoofs, because there is a tradition that its exact location is not published. I'm certainly not going to betray that secret here. But I can hint by saying that I found the wind on my face many times on my way from the village to its roar, something which is very helpful when hunting for secret, sunken waterfalls.

I heard the sound of white water grow as I walked along the road and then crept carefully to the dangerous lip of the hollow. Drawing closer still, I felt the vibrations of the smashing water in the turf below my feet. Below me there was a dark, scooped, rocky cavern, painted green in places with mosses and ferns. The water changed shape as it fell and I would describe it as something of a tiered, cascading waterfall, which fanned a little and had sparkling horsetails, before segmenting and plunging into a punch-bowl. As you can probably tell, I'm not a great believer that waterfalls suit strict categorization.

The pool below the fall spread out into a green and uneven shape. I listened carefully for an effect that I knew would be there but feared may be too subtle to pick out on this occasion. Water doesn't just create sound, it also changes the sounds that travel over it. Sound waves will travel further over water than over land (as do radio waves). This is partly because there are fewer obstacles over water, but there is something else going on, too. The air just above water gets cooled by the water, meaning that the air low down is cooler than the air a little higher up. This is called a temperature inversion and it bends the sound waves back down, creating an amplifier for sounds over water.

It was a struggle to pick up these effects over such a small pool, but it is worth listening for when you are near big lakes, rivers, or the sea. I have a friend who lives on the south bank of a broad part of the Thames in London. He lives slightly closer, as the crow flies, to Chelsea soccer stadium, than he does to Fulham's stadium. But he hears the Fulham matches and not the Chelsea ones, largely because of the prevailing wind direction and the way the sounds of the Fulham crowd travel over the water.

The Chukchi Eskimos know only too well about sound traveling well over water, for they would starve if they didn't. The Chukchi live in the extreme northerly parts of the Russian Federation, at the edge of the Arctic Ocean. When they go hunting for animals like walrus, they watch from high up at first and then take great care as they move, to ensure that no stones touch other stones and that no metal touches metal. If the slightest harsh noise is made, it travels perfectly across miles of the cold air above the ocean and scares off the animals.

Back on land it is worth considering how these pieces start to come together. If you are standing at the edge of a cool lake on a warm day facing into a breeze, with a waterfall a few hundred yards behind you, you may not hear the waterfall at all, maybe only some children playing on the opposite shore of the lake. Stand on the same spot, facing the same way on a winter's day—when the water is often warmer than the air—with the breeze on your back and you are much more likely to hear nothing but the waterfall. Our sound maps will flex with changes in the wind, but also fluctuations in water and air temperature.

A DARK CURIOSITY led me along one of the roads that headed east out of the village. Once well away from the gentle hubbub of the village I listened again, but heard no water. Then I spotted a clump of snowdrops, which surprised me. Snowdrops are very common in gardens and especially in churchyards, but very rare in wilder patches of land. When you come across snowdrops in what appears at first to be a wilder environment, they are nearly always garden escapees and so signal that there is probably some civilization, or has been, nearby. I followed the trail of snowdrops until I found an anomalous dark shape with lines that were too rigid for nature: a ruined building. I paused by the building and listened again, and now I could hear water, the tiniest of trickles, but definitely water. I turned my head and closed my eyes—it is often easier to find the direction of sound sources with your eyes shut, as your eyes will otherwise bully your focus away onto something else.

On my courses, I teach the following method of finding both the direction of sounds and the direction of the wind. Close your eyes, then listen (and in the case of the wind, feel, too), twist your head slowly either way until you are confident of direction, then point at it and only then open your eyes. If you do this, you can be confident that your eyes won't steer your pointing hand toward something more visually convenient, like a nearby tree that stands out.

Soon I was bending down and peering at the weakest trickle of water as it worked its way down the roadside bank, between ivy, bramble, and creeping buttercup. I spent about ten minutes studying this infant stream, looking, listening, touching, and tasting it. In truth I was desperately keen for it to reveal some wonderful clue, but other than the thriving colonies of moss and buttercup that it hosted, it revealed little that I was able to get excited about. I paused, closed my eyes, and listened again. The gentle sounds of the running water were masked by the wind in the trees and the recent raindrops falling down through the leaves, but would reemerge with each lull in the wind.

The exercise made me tune finely into the wind strength and direction. Then I picked up a sound that didn't fit with the noises of either the water or the canopy overhead, although it, too, undulated with the breeze strength and direction; I couldn't make out what it was. Five minutes later, minutes that grew increasingly filled with an angry guttural roar, the source became apparent as two scrambling bikers sped past with grins and nods of appreciation as I leaned into the side bank. Even on the rare occasions when listening to and for water doesn't reveal interesting things about the water itself, it will always, without fail, add a little depth to our sense of what else is going on around us.

My walk took me into a field where I spent a few moments standing by the Riley Graves. The risk of plague contagion meant that church services were held in the open air in Eyam and people were forbidden from burying their dead in the usual graveyard. They were instructed instead to bury their loved ones in open land or their own gardens. On August 3, 1666, the Hancock family's unbelievable torment began as they were struck by the plague and two of the children, John and Elizabeth, died. Four days later, two more children, William and Oner and their father, John, died. Two days after that, another of Mrs. Hancock's children, Alice, died, and the day after that the last surviving child, Anne, also died. Mrs. Hancock had to drag the corpses of her husband and six children into a field, dig their graves and bury them. I stood by the wall that now encircles their graves but soon found myself walking away, overcome.

As the low sun broke through high clouds, I passed one of the "boundary stones," markers placed by the villagers to signal their limits, the outer edge. These stones formed part of an old and terrifying map, one set in the land itself and marked the line that neither villagers nor outsiders should cross for fear of spreading or catching the plague. We make maps for lots of reasons, and there is no rule that dictates that we should carry only one. Every patch of land is enriched by adding to our bundle the rare map formed by listening for water.

# 11

# Reading Waves

THERE ARE NUMEROUS recorded instances of Pacific Island navigators finding their way under overcast skies, often at night, just from the feel of the ocean's waves under the boat. In one instance the navigator was reported to have relied not so much on a gut feel, but by sensing the motion of the ocean in his testicles.

Water waves vary enormously in size, from ripples that are so small that we cannot see them, to waves that take twelve hours to pass and are 12,000 miles from crest to crest. The one thing that all waves have in common is that they take energy from one place to another. This energy can in theory come from anywhere, but in the oceans there are only three main sources: the moon, earthquakes, and the wind. The moon creates the tides, which have their own chapter, and earthquakes can create powerful waves, tsunamis, which we will return to in the Rare and Extraordinary chapter, so here we will focus on by far the most common: wind waves. The wind passes over the water, imparts some of its energy to the water, this energy moves in a direction, and this is visible as a wave.

This idea of waves taking energy from one place to another is important, because it is very tempting to think of waves as the water moving horizontally, but that isn't what is happening. Think of shaking out a bedsheet. A very visible wave takes lots of energy from one end to another as it travels from the force that is giving it the energy, in this case one pair of hands, to another place, in this case a whipping sound at the other end of the sheet. But, the sheet itself hasn't moved horizontally, only up and down. Watch any waves out at sea and your eyes will tend to follow an individual wave as it travels, giving the impression that the water is moving with it, but focus on anything floating on the surface, like seaweed, a piece of wood, or a bird, and you will see how they stay in the same place as the wave's energy moves them up and down, but not along.

If you watch very carefully, you may notice that although the object returns to almost exactly where it started, it does go on a very small orbital journey. As the wave arrives, it first sucks the object back toward it, then lifts it up, then pushes it forward and then lowers it again, a little like the wave is turning a handle. And if we are going to be really persnickety, the motion at the top of the wave is slightly faster than the motion at the bottom, so the object will get moved a tiny bit in the direction that the wave is moving, but so little that this is often barely noticeable.

So, the basics are simple. The wind gives the water some energy, and this energy travels as a visible wave from one place to another. But this leaves a lot of questions unanswered:

Why do we sometimes get huge waves on calm days? If the wave's water is only moving up and down, how come I've been knocked flat by them? If I blow some ripples into the water on a beach in Cornwall, will these reach New York? And to answer these questions and a lot of more, we need to get to know waves much better and in their four stages of life: their birth, their life on the open ocean, their life in the shallow, and their death.

IF WE THINK of waves as creatures, then it helps to know their anatomy. Waves have certain identifiable parts and characteristics. The crest is the highest part, the trough is the lowest part, and the height of a wave is measured from crest to trough. The wave's length is defined as being the distance from one crest to another and the wave's period is the time it takes in seconds from one crest to the next when measured from a fixed point that they are passing.

As soon as we start using terms like wavelength and period there is a risk that we feel the beauty leaching away, or as the ocean scientist Willard Bascom put it, there's a danger that the study of the ocean falls into the hands of those who've never seen the sea. But try to befriend these terms, as they are only labels that will accelerate your ability to read the waves. The period of the waves is probably the least familiar to most, but the most useful in recognizing the different types of waves out there. Blow into your tea and try to work out the period of those ripples—you'll struggle, because it is so short, but it's worth a go just to demonstrate the point.

Baby ripples like these have very short wavelengths, which means a lot of them hit the side of the cup each second, which must mean their period is very small, much less than a second. If you set up a wave moving from one end of the bath to the other and back again then you'd stand a good chance of getting a vague estimate of period, a second or two perhaps. If you are standing on the beach feeling the waves wash over your legs, start counting as one wave passes your legs and stop when the second reaches them. A count of six "elephants" (one elephant, two elephants, etc.), and you are standing in waves with a period of six seconds.

Now look out at a patch of water and find an easy spot to watch waves pass a stationary point, like a buoy in the water. There may be lots of different causes of waves in this environment—steady breezes, sudden gusts, storms a thousand miles away, boats in a harbor, and others. Try to notice how each group of waves not only appears different from each other but has its own wavelengths and periods. You may also spot how the longer the wavelength, the faster the wave travels.

The next thing to notice is how, when a wave travels over a long distance, it will tend to both lose height and segue toward a smoother, gentler appearance. You can demonstrate this effect in a pond easily. If there is a disturbance in the center of the pond then the ripples will spread out from the center. Notice how they are less high near the edges than the center of the pond. The circumference of the wave just after the pebble lands in the water would be very small, let's say five yards, but a few seconds later, as it reaches the edges of the pond, the same wave is spread around a circumference of perhaps fifty yards; the same energy has been spread across a circular wave that is now ten times the size. The energy spreads out and this leads to a wave that is less high.

## BIRTH

If a breeze blows over a calm patch of water, the surface becomes ruffled. If the breeze dies down again, then the ripples die away quickly and the surface of the water returns to tranquility. However, if we are looking out to sea from a beach, watching the waves roll in with the breeze, and then the wind dies down, the waves continue and calm doesn't instantly return to the sea's surface in the way it did with the ripples. Even an hour later, there may appear to be no apparent change to the size or character of the waves we see approaching. The reason for the difference between these two situations is critical to understanding sea waves.

It is best to think of water waves as being one of three types: ripples, waves, or swell. If the circumstances are right, ripples will develop into waves, which may become swell. But the vast majority of ripples will die out long before they reach these second or third stages. It is a little like seeds, seedlings, and mighty trees; many start life, but few are around a long time later.

Watch any body of generally calm water on a day with the occasional breeze and you will see a great example of how short-lived most ripples are. As a gust plucks at the water's surface a rippled area is formed, but seconds later the ripples have disappeared and the water has returned to calm. This is a very common effect and has its own nickname, "cat's paws," as it looks like the wind is pawing at the surface of the water.

We saw in the first chapter how water is held together by a bond between the molecules and how this creates surface tension, strong enough to support insects. This surface tension tugs these ripples back down, flattening them as soon as they have been created. Whenever we see ripples in the surface, we are watching a battle between the surface tension and the breeze; the tension never stops working, which means that as soon as the breeze dies away that tension will flatten the ripples entirely and iron the water's surface smooth again. This is why ripples are evidence of what is happening right now; cat's paws don't reveal anything about the conditions even a minute ago, only at that second. These ripples are also known as "capillary waves."

In small sailing dinghies, cat's paws are vital to understanding sudden local changes in the wind; there is often no other way. If you sail these boats or listen carefully across the water to those that do, you will hear calls such as "Gust on!" as one crewmember tells the helmsman that they've spotted a cat's paw and that the gust is about to hit the boat. In racing this can be an opportunity to capitalize on a sudden, extra burst of energy, but in all small craft these cat's paws act as a premonition, preventing nasty surprises that can rock or capsize a small boat.

AT THE COAST, there is an interesting pattern that forms in the water that will prove that you are looking at ripples, not waves. In light, fairly steady breezes, any large body of water will be covered in small ripples. But since these ripples are only sustained by surface tension, anything that dampens that tension will snuff out the ripples and return the water to glassy smoothness again.

Most coastal waters (and large lakes) have areas where there is a very thin layer of oil covering the water's surface. This is sometimes the result of pollution—all it takes is the tiniest drop of oil from a boat engine to cover a noticeable area—but they also often occur naturally as the result of organic oils from animals and algae. (In fact there is some evidence that organic oils calm ripples much more effectively than industrially processed oils.) These oil layers are very thin indeed, sometimes only a molecule or two thick, and do not indicate a major incident. But they are amazingly effective at dampening ripples, which is why on these light-breeze days, you will see patches of glassy calm interspersed with ripples.

Any oil spreading across an area of rippled water will be clearly visible as a series of lighter areas. The reason that water with oil on it appears lighter in color than the surrounding water is because the ripples have been flattened out, which means that more of the sky gets reflected.

The breeze often corrals the oils, and therefore these glassy areas, into long "slick lines" that stretch quite far. But this effect only occurs among ripples, as once the wind picks up and the ripples become waves, the dampening of the surface tension by the oil no longer calms the water completely and all the glassy areas disappear.

*Slick lines.*

Since all it takes for the ripples to die away is a lull in the breeze, it is a wonder that bigger waves are ever born. For ripples to become waves, we need a constant wind, ideally blowing in a consistent direction for a period of time, typically an hour or more. Calm water doesn't offer much resistance or friction to the wind, it is so smooth, but once there are ripples, the water's surface is much rougher and it catches the wind more easily. So as soon as ripples are formed on a windy day, a self-reinforcing cycle is set up, because it is now so much easier for the wind to gain a purchase on the water's surface.

If a steady wind blows across ripples for long enough, it causes the ripples to undergo a small metamorphosis. They are given enough energy from the wind to become big enough and strong enough to break the bond of surface tension. Remember the surface tension of the water is very strong indeed on a microscopic scale; it will hold very small pieces of metal afloat—but it is not a strong force on a large scale, which is why we can't walk on water. So once the waves have sufficient energy they cease to be dampened down in the way that ripples are. They have survived to adolescence and are known by a new name: gravity waves, because it is no longer surface tension that is governing their fall, but gravity. The key thing is that now that the waves have emerged from their subjugation to surface tension, their energy will travel much farther and for much longer before dissipating. Ripples in a breeze will last seconds once the breeze dies down, but waves at sea can last for several hours with no wind at all.

IF THE WINDS are very strong and blow for long enough, then waves gather enough energy from the wind to reach another stage of maturity and a higher level of energy, which is called "swell." Swell is best thought of as waves that have enough energy to travel well beyond the place of their origin. Ripples will struggle to reach the far side of a pond if the wind dies away, waves will only travel a few miles without a supporting wind, but swell will cross great oceans—thousands of miles is common. Closer to the source, swells are steeper, but as a swell travels long distances its shape will change slightly, becoming slightly flatter and with a less-steep face.

There is a difference in the behavior and appearance between swell and waves, but it is not a scientific cut-off point. This is where we return to the period of waves, because the simplest difference between the three types of waves—ripples, waves, and swell—is the time between each crest passing. A period of up to a second is a ripple, anything close to or over ten seconds is swell, and anything in between is a wave.

Swell is a long-term trend, one pattern often lasting for days at a time, and the other wave types can come into existence on top of the swell, without in any way diminishing it. It is not unusual to have ripples on waves on swell, and it is even in theory possible to have swell marching one way across the ocean, waves going against it, and ripples springing up in another direction altogether. That pattern couldn't last for long, but it could and does happen sporadically.

Once your eye is practiced at spotting ripples and waves, the underlying swell is easier to pick out, as you become able to filter out the other two. Our appreciation of this layering tendency is also important because it is very common for one swell pattern to sit on top of another one or even more than one.

Waves will not ride on top of each other for any serious length of time as they interfere, and the wind driving them will cancel the earlier patterns quite quickly. But swell is different; it will pass through and under everything else, including other swells, largely unaffected, continuing on its way even as it is ridden over by a storm's waves.

This is where the expertise of the Pacific Islanders really comes in. Telling the difference between ripples, waves, and swell is only the start of it, and to the experienced navigators not a difficult challenge, but it is important for them to not just identify that they are dealing with swell, but to identify the individual swell. They do this by judging its shape, its period, and its rhythm. Each swell will have a signature pattern that combines these elements, and sometimes these are easier to judge by sensing the motion, rather than looking at the sea. This is an effective way of filtering out all the ripples, and most of the smaller waves from the motion. A helpful analogy here might be listening for certain sounds in a busy room. I have often noticed how parents of young children can carry on a normal conversation in a room of hollering children, clattering toys, loud music, and a mobile phone ringing all at the same time. This is only possible because of a selective tuning to and focusing on the sound waves that are most relevant at one time. The Micronesian and Polynesian navigators have managed to do this with swell patterns, picking out the meaningful rhythms, when many might struggle to see beyond an anarchic melee on the surface of the sea.

The character of ripples is relatively simple as their lives are so short, but waves will develop according to three main influences: the strength of the wind, the length of time it has been blowing, and the "fetch," which means the distance of open water that the wind has blown over. Each of these needs to be above certain minimums for waves to form and then an increase in any of them will lead to bigger waves.

It is important to note here that the wind never generates perfectly uniform waves; it creates families of waves of similar characteristics, but there will be variability among them. This is why it is the convention for reported wave heights to refer to the average of the highest third, not the highest wave out there. It is popularly supposed that every seventh wave will be bigger than the others, but the truth is all waves are likely to be a very similar size to the others in that set, but there will also be outsiders, waves that rebel. The chances of a wave being a bit bigger or smaller than the one behind it are quite high, but the probability of it being fully double the height of most of the waves you see is very small, approximately 1 in 2000 according to ocean scientists. We will return to the real outliers, "rogue waves," in the Rare and Extraordinary chapter.

We all know instinctively to expect a rough sea on a very windy day, as the wind's speed is the most familiar influence on wave heights. But we can now be a bit less vague about that relationship, thanks to developments that culminated in the work of Rear Admiral Sir Francis Beaufort.

At the start of the nineteenth century, Beaufort, an Irish Naval Officer, must have appreciated two important things about sailors and sea conditions. Firstly, water folk have a tendency to be subjective and exaggerate. From the fisherman's "one that got away" to the sailor's "waves the size of a mountain," there was and always will be a lot of hyperbolic nonsense blowing around. Secondly, and just as importantly, he realized that perfect precision was not the answer to this problem; sailors dislike precision almost as much as they dislike bureaucracy.

## THE BEAUFORT SCALE

| Beaufort wind scale | Average wind speed in knots | Wind description | Probable wave height in feet | Sea state |
|---|---|---|---|---|
| 0 | 0 | Calm | 0 | Glassy |
| 1 | 2 | Light air | 0.3 | Rippled |
| 2 | 5 | Light breeze | 0.7 | Smooth wavelets |
| 3 | 9 | Gentle breeze | 2.0 | Slight |
| 4 | 13 | Moderate breeze | 3.3 | Slight–Moderate |
| 5 | 19 | Fresh breeze | 6.6 | Moderate |
| 6 | 24 | Strong breeze | 9.8 | Rough |
| 7 | 30 | Near gale | 13.1 | Rough–very rough |
| 8 | 37 | Gale | 18.0 | Very rough–High |
| 9 | 44 | Strong gale | 23.0 | High |
| 10 | 52 | Storm | 30.0 | Very high |
| 11 | 60 | Violent storm | 37.7 | Very high |
| 12 | 64+ | Hurricane | 46+ | Phenomenal |

Part of the genius of the scale that was named after Beaufort, although others contributed to it before and after him, is that it fits so well with the nautical mindset—it appreciates that those at sea don't gauge things the way scientists do; they sense and feel things, more like poets. There is something metaphysical about time at sea—I know one nautical professional who is fond of saying, "There is no such thing as an atheist transatlantic sailor." I think it helps explain why the Beaufort Scale has triumphed. It marries science with sensibility.

THE BEAUFORT SCALE works so well because water behavior and wind have such an incestuous relationship. The Scale allowed those at sea to report wind conditions by looking at the sea and deciding which category the conditions best fit. Over the years the Scale's use has evolved and actually been turned back to front—it has now come to be used more as a tool for predicting sea conditions based on weather forecasts, than as a tool for reporting wind strength based on sea observations. Sailors to this day can predict the sea conditions to expect based on a forecasted wind strength presented as a Beaufort Scale number.

The Beaufort Scale refers to conditions on the open ocean and this is important because of one of the other two major factors in the life of wave creation: the fetch. One of the biggest differences between beginner sailors and more experienced ones is that early on it is easy to read too much into the forecasted wind alone. If the forecast is for a Force 5, a new sailor might think, I've been out in a 6 before and it wasn't all that bad, so this will be fine. A wiser sailor will think, where's this

5 coming from? Because the more open ocean that a wind can blow over, the greater the wave size that will result. A Force 5 that has come across hundreds of miles of uninterrupted Atlantic creates a completely different sea from one that is blowing offshore, with only a few hundred yards of sea between you and the wind. Listening to the Shipping Forecast, the sound of a Force 7 forecast in places like the Faroe Islands makes me shiver much more than a Force 9 in Dover, in South East England, does, not because it's farther north, but because it is blowing over open, unprotected sea.

You can witness this effect on a small scale every time you look out across a lake when the wind is blowing from behind you. Just notice how the water closest to your feet is relatively calm, but the water in the distance is forming ripples that grow in size and, if the wind is strong enough and the lake big enough, waves in the distance. Conversely, if the wind is blowing toward you there will be waves lapping at the shore near you, but relative calm on the far side; this is a basic map of the effects of "fetch."

It is also true that the longer the wind has been blowing, the more energy the waves take on and the bigger they become. Given enough time, waves on an open sea will get close to three-quarters the speed of a constant wind. These three wind factors together, strength, fetch, and length, will determine a lot of what you sense over open water. They also help to explain why we will notice certain dependable diurnal patterns, like smaller waves at night—there is a saying, "when the sun goes down, the ocean lies down," which stems from the fact that the sun is the driver of the winds. The sun's energy causes air to warm and rise in some areas more than others, especially over land, leading to pressure and temperature gradients, and these are the main cause of our winds.

A STORM WILL generate swell that will quickly outrun the storm itself, so you should expect to see large swell arriving ahead of a serious weather depression and its storms. The storm will generate waves of different characters, which then spread out ahead of the storm. The waves with the longer period and wavelength will travel fastest and arrive first, followed by the sets with sequentially shorter periods and wavelengths. This means you can gauge the approach of a storm by timing the waves; as the time between each crest decreases, the storm draws nearer. Conversely, if you see some very dark skies approaching from the horizon, but the seas remain relatively calm, it is very unlikely that it is anything more than a local patch of bad weather, a squall that will quickly pass.

In the long era before satellites aided weather forecasters, the behavior of the sea was often the best early-warning system of danger to come. Islanders have traditionally known that the sight of large seas on clear days is not a good omen—it is usually a sign that the swell has outrun the storm that created it, but that the storm cannot be far behind. On September 8, 1900, locals remarked on the extraordinary swell hitting the beach in Galveston, Texas. The next day they were hit by one of the worst hurricanes in US history, one that killed more than 6,000 people.

Surfers are fond of using this logic the other way around. News of storms travels quickly in our electronic age, and surfers know to head to the shore to enjoy the swell generated by the storm, long before the skies darken.

Given the fury at times on the sea's surface, it is hard to imagine that beneath this wet violence, all is tranquil. A submarine need only descend to 150 yards beneath a hurricane to reach calm seas.

## WHEN WATER MEETS LAND

When waves come into contact with a coastline they will typically do three things. Waves will get reflected, refracted, and diffracted. We first saw some of these effects in the Pacific in a Pond chapter, but now it's time to get to know them in a way that means you'll be able to recognize them at pretty much every place where the sea meets the land.

## REFLECTED WAVES

Let's start with the simplest and the one most people are familiar with. When waves meet an obstacle that is anything close to vertical then these waves get reflected, and the steeper the face they meet and the deeper the water, then the more perfectly they get reflected. The next time you see waves being blown onto a vertical cliff, look at how the waves get reflected, as this is a measure of what is happening underwater. If there is a gentle underwater shelf then the waves will break at some point before the cliff and lose nearly all their energy before hitting the cliff, and so they don't get reflected. But if the water is deep all the way to a perfectly vertical cliff, then the waves will get reflected back with almost all their energy.

There are some places where we can watch waves hitting a vertical surface in deep enough water that they bounce back perfectly and one of the best of these to study is a sea wall.

When waves hit a sea wall they lose very little of their energy and will bounce back as waves that are pretty much identical to the ones that hit the wall. This can create some interesting patterns in the water well worth looking for.

If the waves hit the wall square on, then the reflected waves will head back out in exactly the opposite direction that they came from. But the waves will still be arriving, which means we've got waves coming straight at the wall meeting waves heading back out. As these waves pass each other the crests meet each other and the troughs meet each other, forming super-crests and super-troughs, making the waves momentarily appear twice as high and deep. In certain situations this can form a weird and wonderful pattern in the water known as "clapotis," from the French meaning "lapping," which is when the troughs and crests of the incoming and outgoing waves create a standing wave. When this happens, it stops looking as though any waves are arriving or leaving; there just appear to be waves going up and down in the same place and not traveling anywhere. In a clapotis there will be lines where the water is rising and sinking dramatically with a set rhythm and between these there will be lines, known as nodes, where the water barely appears to move at all.

It is quite rare for the waves to hit the sea wall perfectly square on. What is much more common is for them to hit the wall with a glancing blow and then bounce off in the reflected angle, just like light does when it hits the mirror at an angle. This still creates strange and mesmerizing patterns in the water, more of a cross-hatch effect, which is why this much more common phenomenon is known as *clapotis gaufre* or "waffled clapotis."

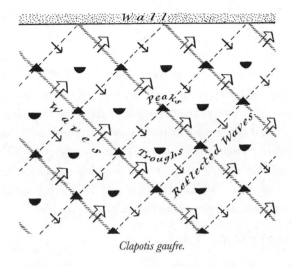

*Clapotis gaufre.*

Don't be disheartened if you don't see a beautiful clapotis or its more popular cousin the clapotis gaufre on your first inspection; I didn't see one worth getting really excited about for a few months after first searching for them. Just bear a couple of things in mind: The steeper the obstacle that waves hit and the deeper the water there, the more truly the waves get reflected. And when reflected waves meet oncoming waves interesting patterns will be created, no two identical, only a handful mesmerizing, but all worth at least a moment's getting to know.

The effect of reflected waves is the reason why the water does some unexpected things on very steep beaches. You may have experienced that strange feeling when you are only in shallow water on a steep beach, but the water seems to be doing a lot of chaotic things around you, crests jumping up, foam leaping toward your face, despite there not being any roughness in the sea further out. Reflections are everywhere; you will even see what look like ripples emanating from reefs if you look out of enough aircraft windows.

It follows that the less the waves get reflected, then the more the energy in these waves is being absorbed by the barrier in question. This is why breakwaters have to be built in relatively deep water. There is no way they could withstand the power of a storm's breaking waves in shallow water. In deep water these big waves don't break and just bounce off the wall harmlessly. When it comes to stopping the power of the sea, it helps enormously if you can pick where the battle will take place.

## REFRACTED WAVES

Whenever anything passes from one area to another and changes speed as a result, there will be situations where it also changes direction noticeably. When waves head toward the coast they will reach an area of shallow water and at this point they start to travel more slowly. It is tempting to think of the waves being slowed by friction of the land, but that is not what is actually happening. It is the effect of shallow water on the waves themselves that slows the progress of the waves. Once the depth of the water is half the wavelength of the waves, it effectively cramps the motion of the waves, and it is this that slows them down.

As we saw earlier, the wavelength and the speed that the waves are traveling at are both reduced—they are slowing down but also getting bunched up, so the period doesn't change. What this means is that if you were on a surfboard a good distance out to sea waiting for the perfect set of waves, you would see fast waves pass under you, with long distances in between each wave. The person paddling in shallow water nearer to the shore will see the same waves arrive, but they will be slower with shorter distances between them. But, if you both timed the number of waves that passed you each minute, you'd get the same answer.

So, shallowing seas mean slowing waves and slowing waves mean a change in direction, and this is where things get interesting. The seabed shape tends to mimic the shape of the coastline, which means that where the land juts out, the sea gets shallower there first, and where there is a bay the water stays deeper for a bit longer. This makes the waves slow down and turn toward the shore near headlands, but they carry on straighter near the middle of bays and then fan out toward the edge of bays. This means that wherever you stand on a coastline, you should expect to see the waves bend toward you. This is also the reason why we get crescent-shaped beaches between two headlands—the waves tend to fan out as they enter any bay and the sand is spread out wide as they do.

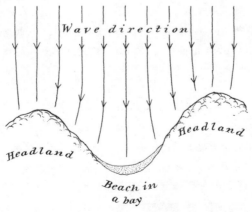

*Refraction causes waves to bend toward land at the coast.*

There are physical laws and formulae to describe this effect, like a fiendishly complex one called Snell's Law, but we don't need laws to make sense of things. It is one of those simple realities: Water gets shallower at the coast and this makes waves bend until they are parallel to the shore. (You might find it easier to remember this effect by thinking of the land having a magnetic attraction for the waves, even if deep down you know that the waves are not being attracted by the land, but change direction as they slow in shallow water.)

Long before the physicists attached labels to the effect, the Ancient Greeks were unsurprisingly all over it. There are references in Homer, Apollonius of Rhodes, and many other sources to the way waves appear to attack headlands, which should not surprise us as nature was busy with these effects a very long time before we started naming laws.

This is one of the reasons that harbors can offer calm and respite for ships. The waves of the open ocean will get bent toward the headlands on either side of a bay and any waves that make it into the bay will be spread out, both effects reducing the force of any waves that do penetrate into a harbor. Or, in Homer's parlance:

> *There on the coast a haven lies, named for Phorcys,*
> *The old god of the deep—with two jutting headlands,*
> *Sheared off at the seaward side but shelving toward the bay,*
> *That break the great waves whipped by the gales outside*
> *So within the harbor ships can ride unmoored*
> *Whenever they come in mooring range of shore.*

In April 1930, a breakwater at Long Beach, California was smashed by waves. This was not the first time that waves have smashed structures like this, of course, and it won't be the last, but this particular assault by the sea annoyed and irritated the oceanographers beyond belief. The problem the ocean scientists had was that according to the science, the sea wasn't rough enough to do this type of damage on the day in question. Not only were the waves not big enough according to all the models, forecasts, and weather data gathered, but also according to observations from offshore gambling ships. These ships positioned outside the breakwater reported surprisingly calm water where they were, even as great stones were being blasted off the structure behind them.

This made no sense to the scientists and they scratched their collective heads for seventeen years until one of them, clearly driven to distraction, had had enough and finally cracked the riddle. In 1947 M. P. O'Brien discovered that there was a hump in the seabed and this hump had slowed and bent some of the waves on that particular day. The shallow water over this hump led to the waves being refracted in a particular and unfortunate way, bending the waves on either side of the hump in toward the breakwater at one exact spot. The underwater bump had created an accidental lens for the waves and focused their energy on that precise location, sending tonnes of stone hurtling.

## DIFFRACTED WAVES

Waves slow and are refracted when they enter shallow water, but something slightly different happens when they encounter narrow gaps. When a wave passes through a narrow

gap—anything comparable to the wavelength of that wave—
then it will get diffracted, which means it spreads out. So any
ocean waves passing through a narrow gap in a barrier will fan
out. Since the energy has to stay the same, but the waves are
now covering a wider area, so the height of the waves will be
reduced overall.

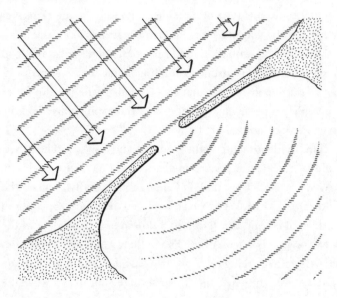

*Wave diffraction through a narrow gap.*

A narrow gap gives the most dramatic demonstration of wave
diffraction, but it's actually happening whenever any waves pass
obstacles. If you hide behind a tree you can still hear a person
speaking on the other side, even though you can't see them,
which is a bit odd when you think about it—that sound isn't going through the trunk of the tree, so how is it reaching you? The

sound waves are getting diffracted, bent around the tree, reaching your ears. Light waves are far too small relative to the tree so don't diffract noticeably, which is why we can't see round trees. But light will get diffracted by much narrower gaps, which is why we see so many colors when looking at a silver DVD.

When water waves pass the end of a sea wall, for example, watch how they don't travel in a straight line, but fan out to fill areas behind the wall, as in the diagram below.

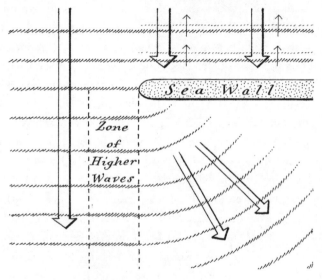

*Wave diffraction past a sea wall.*

Again, the waves are being spread over a wider area, which means their energy is being diluted and the waves become less high as they fan out. However, if you are being eagle-eyed, you may notice that there is one thin zone, directly in line with the end of the wall or other barrier (shown within dotted lines in the diagram above), where the wave height is actually higher than anywhere else, higher even than the original wave before it passed the wall. This is yet one more consideration for the builders of breakwaters; otherwise they can accidentally create more rough water in areas than if they had done nothing.

Waves refract around islands, as we saw in an earlier chapter, but they will also diffract around them, too. Together this means that islands don't provide the shelter from waves that many would predict. My wife, Sophie, and I had a tradition (a telling use of tense there, as will become clear) whereby we would celebrate our wedding anniversary each December by sailing over to the Isle of Wight, going out for a meal, spending the night on the boat and then sailing back to Chichester the following morning.

One Sunday, a few years ago, I knew there was a strong southerly wind, about Force 7, and that this would mean things would be OK as we sailed along just north of the Isle of Wight, heading east back to Chichester, where I live, in South East England. However, for all the reasons above, I was aware that our shelter from this wind and its big waves would not even last until we passed the island. Diffraction and refraction would mean these bigger waves would reach us quite a bit before we passed the eastern edge of the Isle of Wight. I explained this to Sophie and checked that she was happy to come along for what I promised would be an "interesting" few hours.

The knowledge of what would be out there couldn't make those seas go away, though; they only helped to predict them slightly better. Which, as it turned out, was not enough to console my wife as the boat began to be thrown about in what I considered a predicted way and she considered to be something else. She never set foot on the boat again and I sold her a few months afterwards; the boat that is.

## BREAKING WAVES

We have been looking at the behavior of waves in deep and shallowing water, but most of us are more familiar with seeing them when they are in the very shallowest water. When a wave enters water with a depth that is less than half the wavelength, the wave's shape and behavior change slightly. The orbital motion of the wave below the surface gets cramped by the seabed, and this causes the waves to decelerate, bunch up, and the wave faces to steepen. The difference between crests and troughs also becomes more pronounced, with steeper and narrower crests and flatter, broader troughs.

Since shallowing seas can be detected in the behavior of waves long before the waves break or the seabed becomes visible, sailors have used this as a warning sign for centuries. The change can be felt and heard, too, because these waves are harder to sail through than open-ocean waves, and a sensitive sailor may detect a treacley quality to the water, as the boat is slowed, or notice the rhythm or sounds of the boat in the water change. Approaching a coastline this is to be expected and would not cause any surprises, but in an area of scattered reefs, atolls, wrecks, or rocks, any unexpected change in the waves can help focus a wary sailor's mind.

The fifteenth-century Arab navigator Ibn Majid compiled an extraordinary work known to his friends as "The Fawa'id," but with a more unwieldly full title, *Kitab al-Fawa'id fi usul 'ilm al-bahr wa'l-qawa'id*, which translates as, "the book of profitable things concerning the first principles and rules of navigation." It is a hefty tome and one that I am proud to own. In it Ibn Majid refers to the effects above in many places, but my favorite is where he describes an area of choppy water that he appreciates as a sign of shallower water. He returned to the same spot many years later only to discover it had become an island, complete with trees.

As a wave approaches the shore the water gets shallower still and the wave will slow down and steepen further. At this stage, the energy in the water is getting concentrated into smaller areas quite quickly, which causes the waves to grow significantly in height, to rear up. At the same time, the lowest part of the wave slows down more markedly than the top part, which means the crest of the wave begins to overtake the bottom part and the wave starts "breaking."

A breaking wave is the sudden release of stored wind energy, and since the wind gets its energy from the sun, breaking waves are ultimately a release of solar energy. I find it strange and pleasant to think of the sun heating the atmosphere thousands of miles away, winds developing, the sea absorbing the wind energy as waves, the waves transporting this energy to a distant land, and then the boom and roar as the energy is released into the shingle.

Waves break because they have become unstable, and this always happens when the water depth reduces to 1.3 times the height of the waves. But not all breaking waves behave similarly, and the way each wave breaks will be determined by the height of the wave and the nature of the seabed at that critical breaking point.

A debate rumbles on about whether there are three or four types of breaking wave, which is a bit of nonsense really, since you could fairly argue that there are as many types as there are waves. But I find it is helpful to think of there being three families: spilling, plunging, and surging breakers. The general rule is that the shallower the seabed at the point of breaking, the gentler the breaking style. If there is a very shallow gradient, the wave will not break in a sudden violent motion, but instead it falls apart from the top down, sending "spilling" white water cascading down and forward. These are the waves that throw Jacuzzi-like foam up the beach, that don't pack much of a punch and are almost ticklish to swim in.

If the shore is a bit steeper, we will see "plunging" breakers. If you have seen a photograph or painting of a wave that strikes you as a classic, or beautiful in some way, it is almost certainly a plunging breaker. These are the waves that form the most distinct crests and sometimes the "windows" of water in the wave's face that you can see through. They are also the only ones where you can watch the wave form into a distinct breaking shape, the most extreme examples being the "barrels" that surfers dream of.

The third type is known as a "surging" breaker and is usually found only on very steep beaches. Think of a young plunging wave forming, the crest rising up, but then because the beach steepens so quickly, the crest doesn't get a chance to completely overtake the bottom, so the top and bottom ride up the beach together. I like to think of these waves as "slopping" up the shore, as they don't fully break in the way the first two types do. Since they don't break completely and they hit the steep shoreline, these waves reflect back a lot of energy, and the combination of the steep incoming waves and the powerful reflected waves make these shores dangerous for both swimmers and boats; fortunately they are relatively rare and especially rare as destination beaches.

*A plunging breaker, with spindrift.*

As is so often the case at sea, the wind gets to have the final say on how breaking waves behave. At certain wind-speeds the crest of the wave will be whipped off and white spray will be carried off with the wind, an effect called "spindrift." Interestingly it is only normally observed at sea when the wind is Force 8, not lower or higher, and so is used as a way of identifying a Force 8. (On land there is a method, too: A Force 8 will break twigs from trees, but not branches.) An onshore wind will cause waves to break earlier, in deeper water, and increases the likelihood of spilling breakers. Offshore winds cause waves to break later, in shallower water nearer the shore, and increase the chances of a plunging breaker.

If you ever want to gauge the height of breaking waves from a beach there is an easy method that will give you dependable results, even if you are looking at waves a long way off. All you do is walk up or down the beach, away or toward the sea, until the tops of the breakers are in line with the horizon. Then the height of the breakers will be your height plus or minus the difference to the level of the main backwash. In other words, if your feet are still dry, the breakers must be taller than you, and if you would have to get into the water to line the tops up with the horizon, they must be shorter than you.

Waves arrive in groups, known as trains, so it is normal to see a set of waves arrive followed by a period of relative calm and then another train of waves. Since the waves are interacting with each other, as well as the reflected waves, the wind, and currents, this sets up an oscillation in the water's level at the shoreline.

You may have noticed how when you build a sandcastle, complete with robust outer walls, and then wait for the rising tide to swamp your carefully constructed defenses, the sea doesn't knock it down in a very predictable way. The waves may lap at the walls for a minute or two and then recede for a while, then the castle gets assaulted by some very aggressive swash, then the sea retreats right back again for a period, before the castle is finally submerged. This phenomenon of fluctuating levels is known as "surf beat" and is caused by all the wave influences acting on each other and creating this oscillation. It is easy to observe, but very hard to predict exactly what will happen next.

HAVE YOU NOTICED how something floating in the shallowest beach water will be pushed up the beach by the waves, after they have broken? This is very common indeed, but shouldn't happen at all if we think about it, because in theory waves don't move water forward. What's going on? John Scott Russell, a Scottish naval architect, solved the mystery when he discovered a new kind of wave.

The waves we have been looking at are known as oscillating waves. As the water moves to and fro, it is only the energy that moves continuously in one direction. However, Russell was watching a boat being towed in a narrow canal by two horses and when the horses and boat stopped, he noticed that the bulge of water that had been at the bow of the boat became a wave that set off at pace; Russell pursued it on horseback, studying how it behaved. The thing that Russell observed and that surprised physicists when he reported it, was that this was definitely a different type of wave from the ones they had become familiar with, because the water itself was moving, not just the energy. Russell christened it a "wave of translation."

When we see water rushing up a beach after a wave has broken, we are looking at a wave that has metamorphosed from a wave of oscillation to a wave of translation. It is this wave that becomes the swash. These translation waves follow slightly different rules from the normal waves out at sea, which leads to some interesting effects. One big difference being that as these small waves arrive on the beach they are effectively riding on the back of their predecessors, and since, like all other waves, they travel faster in deeper water, you will notice how they have a habit of constantly racing to overtake the shallower wave in front.

So, the next time you are standing on a beach, look for the way foam is carried up the beach toward you on top of the shallowest of waves. You are witnessing a type of wave that was only identified in 1834, the same year that the first practical electric motor was invented.

# 12

# The Omani Delight:
## *An Interlude*

THERE WAS A TOUCH of the John le Carré about my route to the dhow.

My first night in Oman was spent in a simple flat in an isolated coastal village called Qantab. The directions to the flat had not included an address, only a series of numbers in a familiar pattern—a precise latitude and longitude—one navigator showing trust in another. It was the only time I have had to ask a taxi driver to stop for a bit, so that I could get out and look at the stars. On finding the flat, I was sent a text message telling me that someone would meet me there, at some point.

The following day, Will, a young Australian historian of Oman, arrived as I was staring out to sea and informed me that he was going to help me get to the boat. Will used his smattering of Arabic to persuade a kind local in the village to drive us half an hour into the district of Muttrah. There we wended our way through the old souk—"*la shukran, la shukran,*" "no thanks, no thanks"—and sought out the sharp lines and proud flag of the Bait al Baranda building.

Standing in front of the white walls and darker wood panels, we were suddenly surrounded as a traditional dance fizzed into life on all sides. Leaping, dancing, singing, and swordplay failed to distract my attention from a jarring sight and sound amid the melee. At first my brain refused to accept the evidence being presented to my eyes and ears, but on second inspection there could be little denying it. There were bagpipes.

The dancing and piping wound down, uniforms shuffled, lecterns were manned by women, and we were treated to a pair of mercifully short speeches in English and Arabic. Short applause and then the collected Omani dignitaries were led into the exhibition by HH Sayyid Shihab Bin Tarik Al Said, Advisor to His Majesty, the Sultan. I followed them in, at a respectful distance, still lost in thought about Omani bagpipes, or the *habban*, as it transpires it is known locally. (I learned later from an expat that Oman regularly dispatches bagpipers to the Edinburgh Military Tattoo, possibly the most astonishing cultural export I have ever come across.)

For a couple of hours I enjoyed the traditional scenes of Omani life set out in the comfortable surroundings of the well-catered grand gallery—boats in storms, boats in harbor, calm camels, angry camels, sand, sand, sand, and sea.

I was introduced to the Australian artist responsible, David Willis, who has lived in Oman for decades, painting the old way of life. I told David how much I had enjoyed his art, as old video footage of a dhow cutting through waves flickered behind him. Then I loitered among cultured Omani society until I met my planned contact, an always-smiling Omani named Fahad. We walked to a white pickup truck and Fahad then drove us for three hours along main roads and bumpier, dustier tracks to the busy port of Sur. The sun set and then reappeared briefly as the truck hit a rock on the road.

Stepping onto the traditional dhow, *Al Shamilya*, I picked my way among the sleeping, gently groaning bodies and found a piece of unoccupied wooden deck to call home. There was little sleep as the lights and sounds of working boats tinkled around the harbor and kept the gulls vocal. The bright light of the waxing moon poured down from above, a fiery ball raced between Orion's feet, and the mosquitoes partied all night long over my face and around my ears.

A few minutes before five o'clock a silhouette or two responded to the muezzin, the Muslim call to prayer, and stepped across the creaking deck and onto land. Sunrise would follow not long afterward to complete the sleepless night with a burning flourish, but first Jupiter and then Mercury shone brightly.

Over a breakfast of unleavened bread and peanut butter, I met the ragtag crew for this unlikely mission. I had met the American skipper and lynchpin, Eric, in the desert, during a natural navigation course I had run on an earlier visit to Oman. He had lived and worked in Oman for eight years as an Arabic maritime heritage specialist and was joined on the boat by his team of Omani and Indian traditional boat-builders and rope and sail-makers: Fahad, Sajid, Muhammad, Ayaz, and Nasser. This coarse-handed core would be augmented by a small team of Western academics and specialists with slightly softer hands, including a pair of marine archaeologists, Athena and Alessandro, from Denmark and Italy. My job was to bring experience of navigating with the sun, moon, and stars to bear on the voyage. The final member of the team and part-owner of the boat was named Stuart, and he was from the English Midlands.

Stuart had graduated in architecture from the renowned Mackintosh School in Glasgow, where he had met and fallen in love with an Omani girl. They married, he emigrated, converted to Islam, and embraced his new family and lifestyle wholeheartedly. It was Stuart's silhouette that had passed in front of the moon as he headed for predawn prayers at the mosque in Sur.

The sun had barely lifted before the quayside was filled with the bustle of making the boat ready. Worrying winds, evident even in the sheltered harbor, meant a sail change was needed before we could slip and in mid-morning heat, this was a parching and enlightening experience. The dhow had a lateen sail, from the French *latine*, a rig popular since Roman times, consisting of a long yard that supports an equally long triangular sail, one that points toward the bow. You will have seen these sails from time to time, in photographs and perhaps on the water. They are beautiful, nostalgic, and a picturesque, rebellious symbol against the modern obsession with efficiency and progress at all times.

The one thing that was not in short supply in ancient times was manpower. A technology that worked well but needed huge supplies of labor was considered a fair one until relatively recently. It takes one practiced person minutes to change a sail on a modern yacht. Eight of us took two sweat-soaked hours to change the large lateen sail to the medium one. Without any metal involved, every inch or so lashings of coconut fiber rope would be needed, first tying the sail to the long yard and then tying it up with a series of dozens of reef knots. My hands soon felt warm and raw from working the coarse coconut fibers into place.

Jobs on the boat done, I ventured out to buy more supplies of unleavened bread and then walked along a breakwater to inspect the sea itself. There were wind shadows and tidal current shadows aplenty. The harbor was liberally filled with other dhows, some moored alongside jetties, but most at anchor in the bay. I enjoyed noticing that the myriad waterfowl were aligned with the dhows at anchor, all pointing into the southwesterly wind.

Another delay, this time as a replacement impeller for the engine was sought. I passed the time by teaching those who had any interest how we could make a compass with shadows near the middle of the day. As the sun rose toward its highest point, its ever-obedient shadows shortened. On reaching its highest point the shadows reach their shortest. So, I explained, the shadow of this harbor lamppost will get shorter and then longer. If we mark it over the course of the middle of the day, we will be able work out when it was shortest, and this will give us a perfect north–south line.

I have done this simple experiment hundreds of times in many different parts of the world, but I still continue to do it. Partly because it is a relaxing way to pass time, partly because in any group there will always be a few who find it of interest, but mainly because I tend to learn something each time I do it. I don't always learn something new about the sun's behavior, but I do often learn something about how different people relate to the sun and shadows.

Once I traced the curves for the shadow tips for a course I was running for military instructors in Cornwall. The summer one goes this way, I explained as I sketched it, but the winter one bends the opposite way. A hand went up and one of the military survival instructors explained that the way he remembers it is that we are happier in summer than winter: a smile and a frown. I'd never looked at it that way before, but I always do now. Natural navigation takes on another layer of interest with a personal or cultural interpretation of what we see in nature.

On the hot harbor stones of Sur in Oman, I was marking the end of a shadow with some chalk and explained that as we approach midday the change in length of shadow becomes very slight. At that moment, the muezzin call to prayer rang out across the harbor, from many different towers.

"The sun is highest in the sky now," Stuart said, as he watched me mark an X at the end of the shadow.

"Really?" I said. "What makes you say that?"

"That is the midday call to prayer. It is timed to be at the moment the sun is highest in the sky."

Stuart and I looked at each other, then at the marks on the ground and grinned. The pieces came together for both of us as we appreciated what this meant. Whether we are Muslim or not, the midday call to prayer is a cue to look at shadows. All over the world they will form a perfect north–south line. I finished marking the shadow compass with the chalk as a fishing boat landed a proud catch of a small shark and a substantial swordfish.

After the shadows had started to grow noticeably longer, I wandered off to inspect a puddle on a distant part of the quayside. It was far removed from the water's edge and clearly not the result of either sea spray or rain. It rains seriously only a couple of times a year in this part of the world. My best guess was that a catch had been kept fresh with ice in a hold and then unloaded onto a vehicle here, leaving a small pool of freshwater. There was little doubt in my mind that it was freshwater and I had no desire to taste it to prove it. The water and its surroundings were sprinkled with a conglomeration of bird mess, clear signs that the birds had both spotted and rejoiced in this rare source of freshwater. Flies buzzed eagerly.

After a few more hours of preparation, the sun set, the heat of the day ebbed, and we slipped our bow and stern lines. We were under way and under weigh. To weigh anchor means to lift it, to set sail.

The frantic business of setting sail kept all hands busy, and then a tense hour followed as we ducked and weaved our way between small fishing boats and their conspiratorial nets, hidden by the night's darkness. Signs were passed from the bow, via hand signals at the mast to the helm. Buoys, lights, and the silhouetted hulks of dark dhows at anchor glided past our vessel.

The busy times settled into the calmer routine of a watch system on open water. Those who are unfamiliar with sailing often suspect that what they see on a short sail of a few hours is typical of all sailing. But this is a poor representation of typical life aboard any sailing vessel. The one or sometimes two hours after leaving port and before arriving in a new one are usually busy times; sails need setting, warps preparing, fenders put in or out. So any voyage of less than a day gives the impression that sailing is all action. But once clear of port and not near a destination port, a calm settles over a yacht, if the weather allows, which it usually does. As this welcome lull embraced *Al Shamilya* and her crew, I made my way to the bow, where I jotted some notes in my book by moonlight and cradled a sweet tea. I reveled in the sight of Canopus, the second-brightest star in the night sky, to our stern. This is a treat to look out for whenever you travel south as it is a southern star and is hidden from view in northern parts of the world.

Ayaz and Sajid saw me making strange shapes with my hands and joined me up front. They had been told by Eric the reason I was on board and were genuinely curious about what I was up to, performing what may have appeared a strange dance at the bow. I showed them how to find the North Star with Cassiopeia and then how to gauge your latitude using the North Star. It is the same angle above your horizon as your latitude.

I showed Ayaz and Sajid how to gauge this angle approximately with their fists—an extended fist makes an angle close to ten degrees for most people. Then we enjoyed chatting about a famous Arab navigational tool: the *kamal*.

If you fix two sides of any triangle you have a measure of an angle. Stand with your back to the wall of any room and stare at the point where the opposite wall touches the ceiling. This is an angle above the ground; if you measure it roughly by counting the number of extended fists it takes to get from the floor to the ceiling, then you have a measure of a fixed angle. It is five fists from floor to ceiling in the small room I'm writing in at the moment, which means it is approximately fifty degrees from floor to ceiling.

The Arab navigators worked out that an arm and fist aren't bad, but a length of string and wooden board are better. They held the end of the string in their teeth and then measured the height of the stars and sun from the horizon against the board. We had a kamal aboard *Al Shamilya*, but in honesty it is such a simple instrument that there is a kamal of sorts wherever there is a piece of string and lump of wood.

Since the angle of the North Star above your horizon is a measure of your latitude, however you measure it, then if you measure this angle on leaving your home port, you have one good indicator of how to find it on your return. The way this angle was measured improved over the years. The kamal was joined or superseded in different parts of the world by such wonderful tools as back staffs, cross staffs, astrolabes, quadrants, octants, and finally the sextant. The sextant may be a glorious piece of kit and the best ones were very finely engineered and cost a lot of money, but both their ancestry and the logic behind them is unbelievably simple: All they do is measure angles.

I took over the helm. Fortunately the compass lamp was broken, so that was not going to spoil things with its avant-garde glow. But I felt a deep revulsion when an iPhone app was suggested as a replacement compass. I steered between Deneb and Polaris.

The off-watch wriggled into sleeping bags and layered themselves over the gently rolling deck. The sound of the sea was broken occasionally by whispered Arabic and the occasional laugh. Alesandro showed as much disdain for the idea that we could pour a jar of sauce over the pasta he had cooked as I had for the idea of steering by an iPhone app. He chopped and toiled admirably over the midnight stove until we had all dined well at Casa Ale's.

Back up at the bow I looked up as a moth danced above the masthead light, like ash above a bonfire. I watched it for a minute before lowering my gaze to the water. A large fish that looked like a small shark jumped. Except sharks don't normally jump. Then I saw it.

The full moon's light reflected off the water's surface, but it did not do it evenly. Its bright path was broader in the distance, narrower in the foreground, as is usual out at sea, but there was something unusual, too. The beam of white light narrowed very dramatically for a short section in the middle, a skinny waist of calm water. I watched the moon lower and its glitter path on the ocean grow a little thinner. The wind dropped totally and the moon path shrank to a very bright patch of dancing bright light.

With the brightest star, Sirius, and the rest of the Great Dog now astern we sailed on. I held on to my position at the bow for as long as I could; with the danger of fishing tackle ever present even offshore, someone had to be there and I was happy to oblige. Bow lookout is the most feared and dreaded position in bad weather, a vomit-inducing chair of misery for some, but I've always quite liked the foredeck, and in these calm, warm conditions it was bliss.

Nobody forgets their first spell up front in a real blow. Many years ago, I was on the foredeck of a seventy-foot racing yacht in a gale in the English Channel. My foredeck companion, Harry, and I battled to sort a problem with the jib as the waves washed over us with each hard slam of the bow. It was February and we were dressed in many layers, finished off with ocean-worthy oilskins, hats, gloves, lifejackets, and safety lines clipped on. If the boat stayed above the waves most of the time, we knew we'd be fine and we laughed and laughed and laughed. Nervously at first, then breathlessly and finally riotously, spitting sea water out of our mouths with happy cries.

When our watch finished, we slunk like lizards along the quarrelsome deck, back to the cockpit and then down to the galley area. Clear in my memory, I still have the sight of poor souls lining the pushpit, the stern safety rail, retching into the foaming sea. Below decks we stepped past others who were sick or injured and stripped off our gear. We had "eggs" on our tummies. The cold water had filled our boots and soaked every last inch of our bodies, except for some reason it had spared an egg-shaped dry patch, no bigger than my hand, on the last layer before skin over both our stomachs. It was another moment that had us laughing, until the dank warmth of that cabin sent us half-falling asleep, with one arm locked around a pole and a mug of tea still in the other.

A clear view of the horizon, a definite task, adrenaline, and fresh air can all help fend off seasickness in a gale. Harry and I had been served up a violent cocktail of those at the bow. It allowed us happiness on an unhappy ship.

Soon the moon's path was back on the water, stretching away toward the faintest silhouette of mountains in the distance. This time the path wasn't straight; it bent away from the moon, to the north. I was entranced. I had seen these "glitter paths" countless times before, but whenever I notice something new about an old friend like this, it is bewitching. I had never before seen how these paths sometimes curve. I made a mental note, and then scribbled in a notebook to be sure, to investigate this further.

My eyes traced a line down from Orion's sword down to south and noted how close it ran to Canopus. The predawn conditions were idyllic and there was now a social going on at the bow. I showed those gathered a few methods for using the stars. We looked at Orion, Auriga, Cassiopeia, and now the Big Dipper, which had risen in the northeast. The greatest joy came for me in us being able to watch Mercury rise in the east. This very bright planet is always so close to the sun that it escapes notice most of the time. If the sky is clear and you know when to look—it is only ever visible just after sunset or just before sunrise—it can be easy to spot at times. And that morning we were in for a treat.

The sun rose and with it the winds picked up. Those of us off-watch rolled across the deck and then back again. All hands were needed to change our point of sail, which involved working the lateen sail from one side of the mast to the other. Plenty more sweat and our hands grew sorer. Before long the winds grew quarrelsome and darker clouds crowded over us. We were forced to seek the shelter of a bay further up the coastline.

We moored up in a sheltered bay and watched the sky grow darker still. Our hopes of a longer sail had been thwarted by an approaching storm. But I had seen many joyous things and one that I had never noticed before. The way the glitter path curved under that moon must have been before my eyes a thousand times, but I had never appreciated its curve properly before that moment on that dhow on that sea. It is a pattern that I now treasure and one we looked at more closely in the Light and Water chapter. This chapter is about something else, something broader.

I had traveled to the Middle East, keen to step aboard the dhow and study water in a setting that was both novel to me and ancient. I thought I might return with countless riches, an Aladdin's bag brimming with new observational gems. We were forced to take shelter after I had met a lot of old familiar signs in the water, but I observed only one peculiar new thing, that curved shape of the reflections of the moon on the sea.

There was a danger of disappointment, but this voyage reminded me that a water reader's frustrations turn to something closer to delight if we learn to accept that the signs we seek are out there and we will find them, but not always according to our own schedule. I now call this the *Omani Delight*, after that short voyage. Unlike the sweet, unctuous jelly known as Turkish Delight, for me the Omani Delight is the habit water has of showing you the things it wants to and at a time of its own choosing.

My body swayed a little with the old rhythm of the boat as I looked up at the rugged terrain that surrounded me. The air danced above the rocks of the jebel and the winds snatched at a bowl of dust. I may have stepped off the dhow, but the water off the Omani coast had not finished with me.

# 13

## The Coast

THE LENGTH OF ANY COASTLINE is deceptive because it is fractal—the more we zoom in and the more closely we look at it, the longer it gets. If you look at an atlas of the world, the distance from the southwest tip of Cornwall to the southeast tip of Kent will appear to be about 300 miles. But if you followed the coast on foot you could expect to double that distance, as you followed the twists and turns of each bay and cove. And if you were an ant following every jutting ledge or rock, the distance would shoot up to many thousands of miles. It is a mathematical quirk of nature that there is as much coastline as we choose to see.

There is another mathematical quirk when it comes to considering how much we can see at the coast. When looking out to sea we are used to looking for the line that divides sea and sky, the horizon. There is rarely anything sitting perfectly on the horizon for us to focus our eyes on, so this line can appear to be roughly the same distance away regardless of where we view it from, but it's not. The height of our eyes above sea level will have a massive impact on how far we can see out to sea. There is an especially dramatic difference when you are low.

If you are walking on a coastal path and your height above sea level changes from 330 to 410 feet as you climb a small hill, then you will be able to see an extra few miles out to sea, from 22 to 25 miles. But if you were standing by the sea's lapping waves and then climbed a small hill that was 80 feet high, the distance you can see out to sea shoots up from only 3 to more than 11 miles.

It's actually an even more extreme difference than that, because we not only see much farther as we gain a little height, but this means that the area of sea we can make out goes up even more dramatically. Standing on a small, 80-feet-high hill, you can see over ten times more sea area than you can when you look out from standing on a beach. There's a good reason that sailors were sent up to the crow's nest of ships and why radar and other radio transmitters are situated at the top of masts now. The height of lighthouses is critical.

In this chapter we will look at how our understanding and reading of effects at the coast can be super-tuned by learning to appreciate the coastal world better. Visibility, coastal winds, and the shape and natural history of a coastline are all part of a beautiful jigsaw and this chapter is about first noticing and then fitting these pieces together.

## VISIBILITY

Only 0.04 percent of the fresh water on the planet is held in our atmosphere, but it is an attention-seeking fraction as every moment by the sea will be influenced by the clouds and mists, and their impact on visibility. Even a lack of water in the air will be well noted as a day of gorgeous sunshine and often great visibility. In fact good visibility is a clue to more than just low moisture levels. When the air is very stable and the layers are not mixing at all, the lower layer becomes like a stagnant pond and this traps pollution and dust in the lower atmosphere, giving us hazy conditions and poor visibility. So seeing a very long way is a sign of low moisture in the air, air that is mixing, and low pollution. There are some famous cases where extremely good visibility gets noted, like the Calais, France, clock that becomes visible from across the English channel in Dover, via one of the tourist telescopes.

*Thames, Dover. Southeast veering southwest four or five, occasionally six later. Occasional rain. Good, becoming poor in showers.*

The final installment in a shipping forecast on BBC Radio 4— *Good, becoming poor in showers*—refers to the visibility. It is such a vital part of both life at sea and the experience of looking out to sea that if there is a regular coastal spot where you do your sea-watching, it is a very good idea to seek out a series of landmarks that stretch over varying distances and in different directions. Then you can use these as your personal gauge of visibility on the day.

"There'll be a downpour in twelve minutes . . . Aye, a proper soaking we're in for," says the hoary old local, sucking on a blade of grass and apparently staring into infinity. Then it rains, twelve minutes later precisely, and you find yourself standing under the downpour, vexed by the wisdom of this local, who is now long gone. It's not wizardry, it's just a case of knowing that when, for example, the radio mast in the west disappears the showers are inbound.

Coastal areas are especially prone to fog because they can be attacked on either front, from land or sea, and because the coast is where warm air, cool air, and water so often meet. There are a few different types of fog, but there are two main ones well worth knowing about, as once you've pigeonholed a fog as one or the other, you can predict its behavior quite accurately.

The first is a land fog. These are more common at the colder times of year, late autumn and winter in particular, and tend to follow a night of clear skies. If it is a cloudless night in winter, the heat radiates out of the land and rises away, hence the meteorological name for these fogs: "radiation fogs." After the heat has radiated away, the ground is left very cold, and when moist air comes into contact with this ground it will condense and form a fog. However, these fogs depend on still air and will only linger if the air is very calm. Radiation fogs get stirred up and dispatched once the breeze gets up. Pilots are so often hampered by these morning fogs that they learn to predict their dispersal within minutes, just by keeping an eye on the wind speed. At nine knots, there's still a lot of fog hugging the runway, at twelve knots, it's all gone. (A knot is one nautical mile per hour and a nautical mile is 1.15 normal miles.)

The other type of fog worth considering is a very different beast: sea fog, also known as "advection fog." These fogs form mostly over the sea in spring or summer, when warm, moist air blows over a cool sea and the water vapor condenses, forming fog. This fog may look like the radiation, land fog, but it behaves in a very different way, as it will not be deterred by even a strong wind and so can linger on windy days. I remember being caught in strong winds and an advection fog in a small yacht in the English Channel many years ago; it is an unsettling business, being blown about by strong winds but not being able to see anything.

## COASTAL WINDS

After checking the visibility, the next thing to tune carefully into at the coast is the wind, because if height and moisture in the air determine a lot of what you *can* see, then the wind's strength and direction will dictate a lot of what you *do* see in the water. Many sailors use little pieces of fabric, nicknamed "tell-tails," that are tied to the sails and stays (the wires that give the mast stability), to offer a constant visual reminder of what the wind is doing. This awareness is of course vital for sailing a boat safely and efficiently, but it is no less valuable for making sense of the sea's behavior from land. On land there are usually some quite big tell-tails on offer: flags or smoke both making excellent visual clues as to what the wind is doing.

Once you've given some thought to the wind's strength and direction, try to reconcile the direction of the wind with both its character and the shape of the land around you. For many of the ancients, including the Greeks, the character of the wind and the direction it blew from were so intimately associated that their meaning became intertwined. The Greek words for winds and directions were often interchangeable, the *anemoi* being the wind gods, each one holding court from a different cardinal direction. More recently, the Pacific Islanders devised "wind compasses," which were not mysterious instruments, but better thought of as a strong record and understanding of the characteristics of the winds that blew from certain directions. The winds that reached Tonga from the northwest were warm and moist, those from the southwest clear and cold and the southeasterlies came with their own trademark clouds.

This concept of a "wind compass" has caused no end of confusion for Westerners, probably because the word "compass" is so inextricably linked with the idea of a physical instrument for us. But there is no need for confusion; it doesn't need to be any more complex than this: Next time you feel a breeze that seems a lot cooler than you were expecting, check to see if it is coming from anywhere close to north. If you do this enough times, the checking stops and you adopt a more traditional approach of associating the cold temperature of the wind with the direction north, and you have your own, very basic wind compass. You may not need or even want to know the direction, but that is not the point. This is the start of getting to know the wind in its various moods, and this is integral to reading the sea.

The more sensitive to the wind you become, the more likely it is that you will get to know certain coastal wind characters. Some of these pop up all over the world and some are more local; some last for a season and some for an hour or two. The Meltemi winds arrive suddenly, blowing from the north into the Aegean Sea between May and September, sometimes lasting for hours and sometimes for days on end. They can be fierce and are feared and maritime police regularly forbid small boats from setting out if there are signs that the Meltemi are imminent. It is easy to see how such winds came to be viewed as characters by the ancients, often as treacherous and malevolent, but occasionally as friends, too. King Philip II of Macedon used the Meltemi to his advantage, knowing that ships would struggle to sail north into them in the summer, and so he would be freer to wage his wars then.

Let's consider two of the more dependable global characters: the sea breeze and the land breeze. On a warm day, the morning sun warms the land faster than the sea and the air above the land rises and a circulation is set in place, with cool air moving as a breeze from over the sea to the land to fill the gap created by the rising warm air. At night the cycle reverses as the land cools faster than the sea; a land breeze follows, with air flowing the opposite way. On otherwise calm days, these breezes can be the only wind that will be felt.

This is one of the reasons why on a sultry hot and oppressive day, people still seek out the coolness of the coast, a habit that is as strong today as it was when Homer's Nestor and Eurymedon sought out a coastal breeze to cool themselves in the *Iliad*. One of the sure signs that you are dealing with a sea breeze is when the wind appears to be coming from the sea, regardless of which bit of coast you're on that day. If you walk or drive all the way around an island you might be able to face into the wind, looking north, south, east, and west, always out to sea, on the same day.

The next ubiquitous character worth introducing ourselves to is the katabatic wind. Remember how the heat radiates out of the land under clear skies at night; if this happens on the steep slopes of a mountain then we find the same very cool layer of air that we did when the fog formed, but because it is on a slope it refuses to sit still. Cold air is denser than warm air, and so this cold air begins to flow rapidly downhill. It all sounds like a very gentle and rather cute phenomenon, and sometimes it is, but there is a cumulative effect at work that means that if the mountains are big, steep, and cold enough, perhaps with a covering of snow, then the end result can be some decidedly unfriendly winds. In some parts of the Arctic, sudden violent katabatics came to be known as "williwaws," the origin of the word being unclear and unimportant, since the word so perfectly captures the anxiety these winds create.

Next on our tour is a less well-known and very intriguing coastal wind effect that we might see whenever the wind is blowing anything close to parallel to the shore. Whenever a wind comes into contact with the surface of the earth, it is slowed down by friction, the exact amount depending on how rough the surface is. Mountainous land slows the wind a lot, flat land not so much, and over the sea the wind is only slowed a little. This means that there is always a big difference between the friction the wind experiences over any land and the sea. When winds encounter friction and slow down, they "back" in the northern hemisphere, which means that they bend counter-clockwise—a westerly wind may back to become a southwesterly, for example.

Putting these pieces together, what it means is that winds blowing along the shore will experience different levels of friction depending on whether they are just over the land or the water, which in turn means that they will bend different amounts and end up blowing in slightly different directions. If the wind is blowing along the shore and the land is on the wind's left, for example from west to east along a southern coast, these winds split. This is known as coastal divergence. If the winds are blowing the other way, the wind over the land gets bent toward the wind over the sea and they get squashed together; this is known as coastal convergence. If you suspect either of these, look for diverging or converging patterns in the clouds, then in the water.

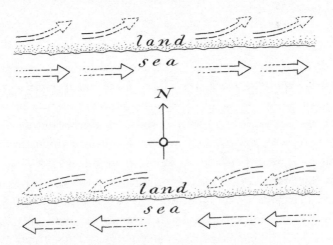

*Converging and diverging winds along a south coast.*

The fact that winds slow over land relative to water also means that you will sometimes notice winds bending to the left as they pass over islands. Nainoa Thompson, a highly respected contemporary Pacific navigator, was able to work out where he was relative to an out-of-sight Hawaii by staying tuned to the way the winds were bent by that island.

Winds bend to the left and decelerate when traveling from water to land, but do the opposite when traveling from land to water, accelerating and bending to the right over water, for example, when they pass over big lakes.

Cold water will slow the wind down more than warm water, and shallow water is often warmer than deep water, so if you look carefully for it you may spot where a breeze bends to the left as it passes over cooler water or to the right over warmer water.

THERE IS A REBEL among the winds over the sea and it is called the "squall." Squalls are complex cells, isolated mini-systems of bad weather that are formed when warm, moist air is drawn upward only to be vomited back down at a nearby place as much cooler, more violent gusts, finished off with a flurry of heavy rain downpours. I will never forget my rumble with these rogues as I crossed the Atlantic single-handed in 2007. Whenever you are sailing across an ocean, not least when you are shorthanded, it is important to keep a really good eye on any weather data and forecasts that you can get your hands on, as well as remaining vigilant for all the natural signs of change.

In the Atlantic you can get hit by some very nasty storms, although the chance of this can be minimized by choosing your departure date carefully. But the truth, one that no professional forecaster likes to admit, is that there is simply no way of predicting exactly where and when squalls will form—they are just too small and whimsical, popping up spontaneously and disappearing as quickly. On my crossing of the Atlantic, I was hit by squalls with exhausting frequency, each one requiring tiring sail changes and something akin to a fairground ride of wind and water commotion for a few minutes, before normality resumed.

However, if you are not responsible for sailing a small boat through a squall, it is much easier to feel friendly toward them, and my grievances against these bandits are now mostly smoothed over, and I enjoy watching squalls from the shore as much as I loathed watching them from a small boat a few years back. The patterns they create in the water can be admired as an angry, slashing artwork might be, and the winds are best thought of as a downward burst, radiating away from under the center of the dark squall's clouds. So, if you see an isolated, angry-looking cloud, make sure you look for any interesting patterns in the water below.

### THE SHAPE OF A COASTLINE

In the 1880s, the Danish naval officer and Arctic explorer Gustav Holm met many Inuit communities as he ventured up the eastern coast of Greenland. He returned with a unique academic souvenir. One of the Inuit he met on his travels, a man called Kunit, sold Holm some lumps of what may have appeared at first to have been driftwood, its harsh edges smoothed by time at sea. But what Holm discovered was that these were very far from random lumps of wood, they were maps. Holm was now the proud owner of three-dimensional wooden maps, shapes that reflected the features of the coastline and could be read by feeling them by day or night. Each bump along the edges of this physical cartography represented a headland or an island, there were notches representing good storage spots, and marks for places where a kayak could be carried over the ground between two fjords.

The coastline where Holm made this discovery is called the Ammassalik coast, and Holm's newly acquired Ammassalik wooden maps made their way safely to a museum in Copenhagen and then, later, back to the Greenland National Museum in Nuuk. There are copies on display in the British Library.

There is nothing stopping us from making our own wooden maps of our favorite coastlines, but there is an exercise we can do that is easier, quicker, and more portable. We can get to know the shape of the coastline by recognizing shapes and then using a trick to remember them.

Thinking back to the phenomenon known as "pareidolia" from the Light and Water chapter, remember that our brains like to recognize shapes and patterns and given half an excuse will invent some where they don't exist. This is sometimes fun when we see faces in fluid scenes like foaming water or clouds—occasionally it is misleading, like when we think dolphins are smiling at us just because we are anthropomorphizing the shape of their mouths. But it is also sometimes helpful, particularly when trying to recognize and remember fixed but complex shapes, like landscapes and, in particular, coastlines.

As you look along a coastline you will see the profile of the land, its shape, a series of promontories, headlands, bays, coves, stacks, jutting rocks, beaches, dips, and all sorts of other landmarks. At first there will be too much information for you to take in, but if you look for the most distinctive features and then allow your brain to play with them a bit, there is a good chance that one rocky outcrop lines up with another and together form a vaguely recognizable shape. This is something you should encourage as it will help you enormously to both recognize and then remember coastal features. This will in turn help you to make sense of some of the patterns that you see in the water.

For example, you might see a large headland in the distance and a more modest one in the foreground that together look vaguely like cats' heads to you. So, you name them "The Pair of Cats." As we have seen, headlands have a huge impact on the patterns we see in the water and forming a relationship with a pair of headlands like this and giving them a name will mean your chances of keeping these landforms in mind and therefore recognizing the patterns in the water that they create will shoot up.

As you move along a coastline, the shapes you see will morph, appear and disappear, so these characters that you create are specific to certain stretches. You have started to create your own mental map of the coastal features around you, and although you won't find them in any museums, these maps are even more portable than the Inuit's Ammassalik wooden maps. Imagine you park your car and take in a coastal view. You spot a rocky feature that reminds you of an old man's facial profile and so you call it "The Old Man." Now you walk along the coast for a bit and within a few minutes The Old Man "disappears," because that rocky cliff no longer resembles him. Later in the day you are walking back toward your car and think you must be very nearly there, but then you think to glance back toward that familiar rock feature: You see no Old Man, so you know you've got a little way to go. A few minutes later you glance back again and The Old Man has reappeared, so you are now confident that the car is not far, and then it appears around the next twist in the path.

This may sound like harmless and perhaps useless fun and games, but it is an introduction to one of the most powerful coastal techniques ever devised by sailors, known formally as using a "transit," and it is one worth getting to know better. Put simply, if any two objects line up, one in front of another, it follows that we are somewhere on the straight line that travels through those two objects. Let's say we see a radio mast on a distant high hill, exactly above a church's spire in the foreground; it follows that we must be somewhere on that line. This is an incredibly powerful way of lining yourself up; it needs no electricity, is easy to use and understand, and is very accurate indeed if you have identified both the objects and can see them clearly.

Les Écréhous are a navigationally challenging group of rocks northeast of Jersey in the Channel Islands. There are a few huts on the main island that we like to camp on in the summer, but to get there you have to negotiate waters with huge tidal ranges and very fast tidal currents, countless out-lying rocks, and history littered with fatal shipwrecks. It is too small for any high-tech navigational aids, and even GPS is of limited use once you're close in as things happen too fast, but the trick that has worked for centuries is to line things up, to use a transit. There are three rocks that jut out of the sea, and all you need to do is line up the one that looks to me a bit like a shark's fin in between the two others and you know you are on the right course. Then you make a turn and line up a black panel on a flagstaff with a rock that has been painted white and continue your course safely to the small island.

For centuries any marks that make the land easier to read from the sea have saved lives, whether they are natural features or artificial ones. Elizabeth I was painfully aware that the well-being of the nation was growing increasingly dependent on the well-being of her seafarers and so one of her less famous, but very pragmatic, decrees was that these coastal features were to be protected: Destroying or changing them was made a criminal offense. To this day, the vast majority of port and harbor authorities are trying to make these transits easy to spot and the colors black and white are often used.

It is well worth keeping this in mind as you explore the coast, because if you spot something conspicuous that is painted black or white, often high up or otherwise standing out, you are very possibly looking at part of a transit. Finding the other part will therefore reveal the line that ships are using to find their way safely in. Of course, glancing at a chart is an easy way of solving any mysteries—transits will appear as a thin, straight black line, running from land out to sea.

The academic and sailor David Lewis was on board a traditional canoe that was sailing away from the Puluwat Islands in the Pacific. He was studying how the local navigator Hipour was managing to hold his course as they sailed through the swell without any instruments to help them. As Lewis suspected, Hipour was managing to navigate by using a transit—he was sailing so that the two islands behind them neither separated completely nor overlapped too much. He would steer so that they were, as he put it, *parafungen*. As Hipour explained this to Lewis, there was a lot of laughing among the Pacific crewmembers. Lewis guessed correctly that *parafungen* did mean islands that just overlapped, but it was a metaphor and really implied two people getting intimate.

These two related techniques of recognizing features and lining up transits are fundamental to all nautical traditions all over the world. They are just as useful and interesting when you are still on land, as a way of recognizing and remembering nearby coastal features, but also of building a deeper understanding of where you are relative to these features.

Another useful method is by gauging the height of distant objects or the angle between them using the most basic of all sextants, an outstretched fist. Last year I parked near some sand dunes and made my way onto Talacre beach near Point of Ayr, the northernmost point in mainland Wales. I began a walk toward the impressive, albeit inactive, lighthouse, but I had barely taken a couple of steps when it struck me that finding the car again might be a small challenge. It was parked in a nondescript place, hidden behind a range of dunes that stretched for miles. I decided to gauge the height of the distant lighthouse by stretching out my fist and counting how many knuckles it was from the beach to the top of the lighthouse. A couple of hours later I was heading back to the car and knew that when the lighthouse had shrunk back to two knuckles, I could pass through the dunes and should be able to spot the car.

If you try this, you'll be following in the footsteps of coastal navigators for millennia. You can either try it by gauging the height of things: lighthouses, headlands, churches, anything with some height will do. Or, you can use the same trick for the horizontal angle between things—maybe the sea end of a pier is an extended fist's width from the land end.

*A glitter path off Lefkada, Greece. It narrows where the "dead wakes"
of passing boats have calmed the sea.*

*Cat's paws on the Blue Pool in Dorset in South West England.*

*Diffraction and refraction cause waves to fan out in Chapman's Pool.*
*There are also interesting color changes, including cloud shadows.*

Al Shamilya, *waiting out a storm behind a breakwater in Oman.*

*Asymmetric sand ripples. In this picture the water has flowed from right to left.*

*Flat-topped sand ripples. Water has flowed from left to right
and then back again with the tide.*

*Symmetrical sand ripples by the Point of Ayr lighthouse in Wales,
formed under breaking waves.*

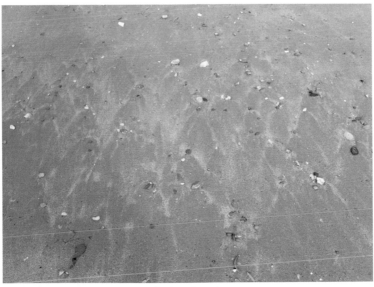

*Backwash marks.*

*Rill marks.*

*Ripples have formed in sand in the trough, where water has flowed parallel to the beach. The bar is on the left and the shingle to the right marks the step.*

*Wake patterns and slick lines in a busy harbor.*

*A speedboat creates a typical wake pattern. But look closely near the top left and you will see the tracks of boats that passed much earlier.*

*The author, researching Viking methods north of Iceland.*

*Long-finned pilot whales off Iceland. Even their gentle wakes change the patterns of light and dark.*

These techniques brought together—recognizing features, spotting transits, measuring angles, and watching how they change as you move around a coastline—have saved countless lives over the centuries, but the point here is less about safety and more about awareness. It is all too easy to see a coastline as pretty without registering any of the rich detail around you, and if you want to make sense of the patterns in the coastal water, you first have to tune in to the intricacy of the land that the water is working around.

## COASTAL CREATURES

When the author Stephen Thomas was researching Micronesian methods of navigation, he was intrigued by a method described to him as *pookof*. Certain birds or fish return to the same feeding grounds on a dependable daily basis, and indigenous navigators, like Mau Pialug, learned to read where land was relative to the more dependable creatures. Every animal has both its habitat and its habits, so it is not a great stretch to imagine these navigators making a map with the help of animals, but it was the detail that surprised Thomas. The Micronesian navigators were not using the broadbrush approach that we might—seeing land-based birds like crows must mean that land is not far-off, for example. They were citing very specific individual features on precise animals: One waypoint location known as *Innamowar* was described to Thomas as being where you see the ray with a red spot behind its eyes.

I'll admit to finding this level of detail far-fetched when I first encountered it. I arrogantly suspected that workable methods were getting diluted by folklore or superstition. But my view changed in unlikely circumstances. I was on a family holiday, visiting my brother's family in Greece, when a random exploration of the Peloponnesian coastline in our rented jeep uncovered a beautiful and quiet beach. We were delighted by our discovery and planned to return the following day. But, embarrassing as it is to admit, it took us far longer to find it the second time than it should have.

We returned to the same beach the following day, too, only this time we found it much more quickly, and we returned again every day for the rest of our week in Greece, finding it easily now with the help of a strange technique. It became known to us as our pookof beach, for the simple reason that the way we ended up finding it easily was to drive along a road, one lined with a large number of identical municipal trash cans, until we saw the one that the black and white kittens were always playing on, and then turn down that small dusty track, which led straight to the beach. The kittens never failed us.

Since my personal and slightly surreal encounter with this method, I have always revelled in finding examples of this intimate reading of the relationship between animals and place. There is a charming thirteenth-century account of this technique being used in the Arabian Sea. "If a traveler in this Sea sees seven birds right out to sea, he knows he is opposite the island of Socotra. Any who travel in this sea and come across the island, will see the seven birds, night and day, morning and evening. From whatever direction ships approach, the birds received them."

All animals will reveal something; it is up to us how much we choose to decipher of their message. There will be lots of broad clues, especially about changing weather, ones that we can take with us anywhere—coastal birds, like gulls, will tend to head inland if bad weather is imminent. But the longer we spend studying one patch, the deeper our reading of the animal clues will become. We might notice some birds that like to climb higher on the thermals, and after several seasons enjoying this, it strikes us that they soar over the sea in winter and land in summer, as the sea is warmer than the land in winter, but cooler in summer. When Robert Stevenson was battling to build one of his famous lighthouses at the Bell Rock, he became tuned to the clues the animals would offer him about deteriorating weather, as the fish would congregate over the reef in good weather, but scatter as bad weather approached.

## OVERFALLS AND FOOLISHNESS

I had five days off work and wanted to spend time at sea, so the plan, if such a vague notion could be called a plan, was to sail west-southwest from Chichester Harbour for about two and a half days and then sail back home for two and a half days. My thinking was that this would guarantee the best value imaginable for the time available—we would get a full five days at sea, instead of hopping from one marina to the next and we wouldn't be able to spend any money at all. The friend who agreed to join me on this exercise of dubious value is named Will, and I remember the look on his face when the trouble that we had invited arrived.

Gordon Tullock, an American economist who died in 2014, became infamous for suggesting that road safety might be improved if we all had a spike sticking out of our steering wheel and pointed at our heart. I believe his point was that things that are designed to make us safer can become counter-productive if they inadvertently lead to us behaving in a riskier way, as a result of believing ourselves to be protected. I can't say whether this theory had any bearing on the short voyage Will and I undertook. It is true that the small yacht I owned at the time had been designed to be about as sea-proof as a thirty-two-foot boat could be, so this may have had some bearing on the subsequent decisions.

The sensible mantra is that if you see squiggles on the chart you avoid sailing through those waters, especially if there are any serious tidal currents running. But Will and I felt that this might not apply to all vessels, and if it didn't apply to all vessels, then it wouldn't apply to ours on the day in question. The squiggles, or wavy lines, on the chart off the promontory called Portland Bill in southern England represented "overfalls," a name given to a phenomenon in the sea when fast tidal currents run over very rough ground, giving rise to turbulent and potentially dangerous water at the surface. The physics are fairly straightforward and can be demonstrated easily at home: If you turn on the tap and let the water run over a smooth, flat object, like a tray, then the water surface stays smooth. But if the same water is made to run over something as relatively rough as seabed caves, holes, and giant boulders, like, say, the coarsest side of a grater, then the surface of the water becomes seriously perturbed.

Mugs, bowls, cutlery, and books spilled out of cupboards and slid around on the floor with an alarming cacophony. We clipped on our safety lines, as the bow of the boat rose up and down in a violent and seemingly unnatural rhythm.

White foam swept over the foredeck as our knuckles turned white from gripping and another surge of belongings smashed onto the floor below decks. This continued for about half an exhausting and alarming hour until we were clear of the over-falls, and then a welcome calm, rolling motion returned to the boat. There had been no fixed aim for this experiment, but we considered it to have been a terrifying success.

I recalled one moment of that experience especially clearly glancing up at the cliffs during our passage through the over-falls, I spotted what appeared to be a coastguard lookout station perched at the top. I imagined a pair of coastguard officers peering down at us, sharing a pair of binoculars and taking turns to shout expletives about what a pair of idiots were in the water below them.

Last year I was in Dorset for work and had a spare few hours and felt it might be satisfying to look down on that same patch of turbulent water from a safer vantage point. I learned that there was a lookout station belonging to the National Coast-watch Institution just up the road at a place called St Alban's Head. The Coastwatch Institution is a volunteer organization that exists to promote safety at sea, and they do this by keeping a lookout. (You'd be forgiven for believing that this is what the Coastguard would be doing here, but no! Her Majesty's Coast-guard now operates radios and electronic devices from an industrial estate in Hampshire, in southern England, that has no view of the sea at all—one more sign of the times perhaps.)

The lookout station was perched on cliffs that overlooked an area of overfalls very similar to the ones that I had enjoyed at nearby Portland Bill. There was a strong wind blowing and

I kept away from the edge; some have fallen to their death on that stretch of coastal path, and the thought occurred that it would be a strange twist of fate, if my desire to see the overfalls from the safety of land led to me being blown off the cliff.

I met the cordial crew of the station and we chatted between the familiar blurts of noise from the VHF radio. I looked out from the broad window and down at the raging white patch that was the overfalls. The sea boiled and hissed and spat foam up into the air. I learned how most yachts that ran into trouble were dismasted by the violence of the motion of the sea in that patch, and I was shown a photo of one such vessel being towed to safety. Listening to the wind whipping around the exposed building, I asked what they thought the wind was gusting to that day. One of the crew glanced at the flag and said, "End of the flag is bouncing up, so we must be looking at 40 knots or more." I loved hearing this—all the more since there were two electronic instruments right next to us. They gave a reading of 41 knots.

There were some distinct eddies as the water flowed past some rocks. Just as eddies will form in rivers and streams as water flows past any obstruction, exactly the same thing will happen in the sea as tidal currents push water past any part of the coast that juts out, from great headlands to small rocks. Then I focused my attention on an area of water close to the foot of some cliffs, one that was behaving noticeably differently from anything else nearby. This was the meeting point of two eddies, a water one and a wind one.

Outside the wind was whipping in from the sea, but only about fifty yards inland there was an offshore wind visible in some flags on nearby cottages. This wind eddy was being created by the cliffs and was forging its own patterns in the water below, adding its influence to water that was being shaped by the tidal current eddy. The only mild disappointment was that there were no skippers stupid enough to attempt sailing past St. Alban's Head that day.

Before I left the station, one of the crew pointed to a spot in the water. "Yachts stop dead right there. There's a hole below that sea that is 53 meters deep."

I gazed down at a small patch of calm white water among the rough crests, one that looked as though it had been ironed.

Soon afterward a red-faced man with shock-white hair, and bloodshot eyes with strikingly light-blue irises burst into the station. He puffed and panted about a cow emergency; about forty Holsteins had made a break through some weak old fencing and were now perilously close to the cliff edge and violent death. The farmer must be made aware of the situation! In light of the drama and sensing that there was little I could do to help the cows, I thanked the crew, made a donation, and battled back out into the wind. A blue Land Rover was rocking in the breeze—and I reasoned, with a smile, that made it a 45-knot wind now.

Secretly I had hoped to watch some younger fool make their way through the overfalls below the cliffs, the way I had years before. But once more the water chose to show me something else. I walked down a path until the coast opened up before me into an idyllic cove, one I remembered well from previous visits to the area, Chapman's Pool. And there the first thing to strike me after admiring the beautiful light-blue color of the water was a classic demonstration of the way waves react to coast-lines, as the headland and bay were being tickled by reflected, refracted, and diffracted waves. In the bay, the waves fanned out and rolled onto the wide crescent beach.

# 14

# The Beach

IN THE 1990s the US military undertook one of the most thorough studies of beach formation processes ever made. It became known as "Sandyduck '97" and took our ability to read beaches to a new level. We may not be planning an amphibious landing anytime soon, but if we take the time to look then we can expect each beach to assault our senses.

The word beach is what we use to describe a collection of familiar features, features that we don't usually notice individually. Every time a wave washes up on the beach it moves some of the sand, then when the water retreats it moves the sand again. The cumulative effect of this happening thousands of times each day creates shapes, and we can use these shapes to read back to what the water has been doing.

We are all familiar with the idea that beaches are steep in some places and flatter in others and we've all experienced that strange sensation when paddling out into the sea of finding ourselves surprised by the depth. These changes in gradient are part of a beach's map and clues as to what the water has been up to. A typical sandy beach will have as many as six zones that we can identify: the dune, the foredune, the berm, the beach face, the trough, and the bar. If you look at the diagram below, you'll see how these features fit together.

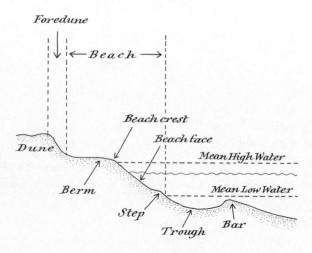

*Classic beach topography.*

The first thing to notice is that the beach doesn't have the same gradient all the way down into the sea; there is a noticeable steepening at a point known as the "beach crest." It is quite a bit steeper on the seaward side of this crest, and this steeper zone is known as the "beach face" (not to be confused with that look you get after a day of too much sun, sea, sand, and wind). On the land side of the beach crest is the broader, flatter area of sand known as the berm—the berm is where we lay our towels, the beach face is where we get surprised by how quickly the water is getting deeper. Of course, the state of the tide will have a huge impact on how much of the beach face is covered; if it is a particularly high tide you may not see the beach face at all, but you will feel it as soon as you get into the water.

If you look out from the beach toward the sea, spot the zone where most of the bigger waves are breaking. Under this area of breaking waves there will be a dramatic shallowing of the water, because this is where the "bar" forms. Wherever there is a sandy beach with waves of any size at all, then a bar has to form, as physicists dictate it, from behind their wave tank experiments. On the land side of the bar there will be a trough of deeper water. On some beaches more than one bar will form and then you get steep undulations of bars and troughs in the seabed. Because of the wave action that shapes and creates these bars, the beach side of the bar is often quite steep, whereas the seaward side is more of a gradual slope. At some stages of tide, if you head far enough out, past the trough all the way to the bar, then this steep side of the bar has a habit of tripping you up.

On a beach I visit regularly, West Wittering in West Sussex, in the south of England, there is always a procession of people expressing surprise as the water depth fluctuates from knee height to waist height and back again, and it is an accepted sport on this beach to trip, try to look cool about it, then watch those who laughed at you doing exactly the same thing. These zones of water on either side of bars often lead to those marked fluctuations in temperature you get, too—patches of bizarrely warm water only a few yards from a much colder sea.

Next time you walk out into the sea, see if you can feel the sand change to a coarser grain underfoot, just before the water gets deeper. There is an area, known as "the step," where the beach face flattens out as it meets the trough area, and there is often a line of coarser sediments here.

As the water falls with gravity back toward the sea, it is getting held up by the bar. This means that the water in the trough is constantly striving to find ways back out to the ocean and freedom. Sometimes this can make the water flow along the beach, parallel to it. I'm sure you know those beaches where you have to constantly swim one way to avoid being dragged sideways. This is very common because the trough is an otherwise lovely place to be, inshore of any big breakers but deep enough for a good swim.

Occasionally the water in the trough makes a break for it and forces a gap to open in the sandbar. Imagine all that water suddenly finding a narrow escape route. You may have already guessed what the consequence will be: a rip current.

In 1998 I was on holiday with my wife, then girlfriend, in Bali, Indonesia. We had been enjoying a swim among the big surf of a beach near Kuta. I was just getting out of the water to join Sophie on the beach when I heard an unnatural sound. Turning around I could see that there was someone on the beach screaming and pointing at someone in the water who was clearly in great distress, arms flailing and a face that was contorted in panic. I looked around for a sign of any lifeguards, rings, or similar, and saw nothing nearby. I was reasonably fit at the time and a fairly strong swimmer, so I felt I ought to do something. I waded back out into the surf and began to swim toward the man in trouble. A hundred things went through my mind during the first ten strokes and I swam hard to beat the incoming waves.

I knew that summoning help would have been a more sensible strategy and I was familiar with the dangers of trying to save someone who is drowning—the psychology of it is bizarre and terrifying. Drowning people sometimes drown their rescuers in their mistaken and often violent bid to help themselves; this is one of the reasons why so many life-saving devices have long bits of rope attached to them. But knowing these things didn't override my gut desire to help. I swam harder and ducked down as a couple of big waves surged over me. Then I surfaced and heard another scary, unnatural, but strangely familiar sound. I stopped swimming and looked around.

The sound was Sophie screaming at me with all her lungs to swim back to shore; I remember clearly thinking that she was further away than I was expecting. I looked in the opposite direction toward the man I was supposed to be saving and the distance between us had shot up, even though I had been swimming hard toward him and he wasn't swimming anywhere. It suddenly dawned on me, better late than never, that we were both caught in a rip current. Fighting an urge to panic myself, I turned and swam hard and as fast as I could straight toward the shore. This is where the psychology of these situations gets really weird. The rational part of my brain was saying calmly, "Don't swim straight toward the shore, you are caught in a rip, you won't beat it that way, you have to swim parallel to the shore until you are out of the rip, then head to shore." But the emotional part of my brain was screaming, "You're about to drown! Look, there's some nice, firm beach ahead of you, swim at that, you fool!"

Since I clearly survived, you may be expecting me to recount how the calm voice of reason prevailed. That is not what happened at all and there is no point in this little story if I am not honest. What actually happened was that the screaming match in my head continued and a truly surreal compromise ensued. I just couldn't bring myself to swim parallel to the shore. Even though the few rational thoughts I had left were that this was what I ought to do, I was just too terrified and wanted so badly to be back on land.

But equally I could see for myself that swimming straight toward the beach was proving counterproductive and dangerously tiring, and so I turned to swim diagonally toward the shore. I staggered out of the shallow surf, a long way from where I had entered the sea, agonizingly exhausted, and then collapsed on the beach next to Sophie. During the following couple of minutes she found a compromise of her own, by simultaneously expressing some relief at my survival and chastising me for being so unbelievably stupid in the first place.

A few more minutes passed before a local came up and crouched next to me as I was still lying flat on my back. He spoke in hushed tones, a conspiratorial whisper, "Thanks for trying, but don't ever do that again. You'll die. We always lose someone at this time of year." He looked out to sea.

"Did the other guy make it back?" I asked.

"I saw some surfers heading toward him, so . . . probably."

Despite trying, I never found out what happened to him.

Rip currents are feared and much misunderstood. They get called "rip tides" frequently, but they are not a tidal phenomenon, and this is just the start of the confusion that they seem to create. The physics is actually very straightforward: rip currents form when a wide area of water on the beach is pulled back to the sea by gravity, but finds a narrow channel to flow through. Water will accelerate whenever it passes through a narrower gap—like when you put your thumb over a tap or the end of a hose—and so a fast current that flows out toward the sea is formed. They can flow at eight feet per second, faster than any swimmer.

Some of these narrow channels are permanent—where there is a gap between reefs, for example—and these rips tend to be well known, by locals at least. But some are temporary, as the returning water surges through a new channel between sandbanks, and these catch more people by surprise. Ironically, since rip currents can have a smoothing effect on the waves, they can attract swimmers who think they are entering a benign and relatively calm patch in an otherwise choppy sea.

Rip currents can be hard to spot from the shore and harder still from water. The only broad rule is that we know that water that is behaving differently will appear different. In the case of a rip current look for a thin channel of water that is either more choppy or less choppy than either side (this will vary depending on where the wind is coming from), a line of water with more foam on it that is heading out to sea, a consistent disruption in the wave patterns along one particular line, or any other anomalous patterns that form a line perpendicular to the shore. Most beachgoers fail to spot any of these, but a water reader can aim to do better. Whatever you do, try not to follow the poor example I set in Bali of both failing to spot the rip and then leaping in and swimming out to sea in it.

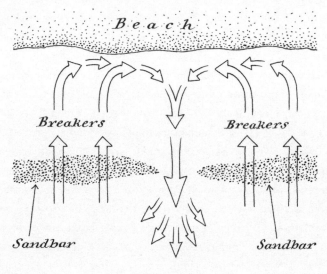

*Rip current.*

It is tempting to think of each of these beach features as permanent, but the whole beach is in a constant state of flux. Bars, troughs, faces, and berms are all being resculpted, effectively destroyed and recreated each year. There will be seasonal changes at every beach, too. Berms tend to be higher and thinner in winter and the bars are significantly bigger then, too, as the more powerful winter waves borrow sand from the berm and build bigger bars.

I subscribe to an email newsletter known as a "Notice to Mariners," which gives me nautical updates for my local area. The one I received on Christmas Eve last year was fairly typical, if a little un-festive, for the time of year:

Mariners are advised that a bathymetric survey of Chichester Bar undertaken 15 December 2014 shows an isolated spot height of 0.9m below Chart Datum close to the western edge of the channel, north of Chichester Bar Beacon. For greatest depths keep east of the line between Chichester Bar Beacon and Eastoke Buoy where the least depth is 1.3m below Chart Datum.

In other words, the bar had shifted and grown in places, as is its seasonal habit. Great storms will do more serious landscaping, rearranging not just the berm and the bar, but the whole beach. Orkney is a place that gets its share of strong winds, but in the winter of 1850 a particularly violent storm battered the largest island, and the onslaught was compounded by a very high tide that day. When things calmed, the islanders were surprised to find the outline of stone buildings in among the dunes, in an area then known as Skerrabra. Today it is known as the UNESCO World Heritage Listed site, Scara Brae, a Neolithic settlement that offers a unique insight into life 5,000 years ago.

Very substantial bars can form at the mouths of rivers where they meet the sea. If they are not dredged then these bars can block the flow of the river's water out to sea and lead to the river "kicking" ninety degrees one way or the other.

## THE ART OF UNDERTOW

It is time to peer closer and look for some of my favorite features on the beach. One of the most familiar sensations on beaches is the wave that washes brusquely past your legs, followed soon after by a gentle foamy tugging at your ankles and tickling of your feet as this water rushes back down to the sea. The water that surges up the beach after the wave has broken is known as the swash and the returning water is the backwash, which then becomes the undertow. The undertow is a flat sheet of water that slides in under the incoming waves. It can be surprisingly strong, which makes it especially ticklish on the feet, but it is never and can never be a "rip."

This is one of the most common misunderstandings; whenever anyone feels that strong tug at their ankles, there is a temptation to start gibbering about rip currents, but the two phenomena are not related. The undertow may feel strong at times, but it is very low and peters out almost as soon as it reaches the incoming waves. Rip currents can carry you out for hundreds of yards, but undertow is rarely a threat to anyone who can walk or swim unassisted.

Undertow may not be a great danger, but it is fascinating and creative. Every time water flows over sand, it shapes that sand, so we can look for patterns and use these patterns to gain an insight into what the water has been up to. At low tide, it is well worth getting close to the sea and looking at the beach. We should expect to see ripples in the sand, many of them parallel with the beach and the lines of waves that have been breaking over it a few hours earlier. But if you look carefully in the trough area, you may spot sand ripples that don't fit this simple pattern.

If there has been a flow along the trough—the one we find ourselves swimming against—then there will be evidence of it in the alignment of these ripples. In the trough, the sand is being shaped by both the action of the waves, creating some ripples that will be parallel to the shore, but also by the current that is flowing along the shore in the trough, creating ripples that are perpendicular to the beach. If both of these ripples are formed then you get an effect called "ladder-back" ripples in the sand, a sort of cross-hatching. Because the current in the trough will be much narrower than the waves going over it, the current-ripples tend to be the narrower of the two.

Bear in mind that all sediment ripples that have been created by something flowing over them follow a simple rule—the side that the flow is coming from will have a shallower gradient and the side the water is flowing toward will be steeper. This is just as true for wind-shaped dunes in the desert and snow ripples in the mountains as it is for sand ripples on the beach their shape is a clue to the direction that something has been flowing. (Look among any sand dunes you find higher up the beach and you will see wind-shaped ripples there, too; try touching either side of these small ripples and you should be able to sense how one side is softer than the other—this will be the downwind side. If you know what direction the wind has been blowing from, then these ripples will act as a compass for you. This is a trick used by the Tuareg in the Sahara, but one that works just as well on any beach.)

If water flowing one way creates a ripple pattern with a steeper side in the direction it is flowing, then it follows that if the ripples are equally steep on each side this must be a clue to something else. These symmetrical sand ripples are the result of water that is oscillating, flowing both ways over the sand, and are very common near waves that are breaking. However, if the water isn't oscillating but has been flowing one way for a while then in the opposite direction—very common in places with big tidal ranges—then the sand ripples will have distinctly flat tops, as the normal crest shape is first formed and then sliced off and ironed flat as the water heads back against the grain.

If waves have been coming from two different directions, as happens when there is change in the weather and therefore wind direction, then this will create a different pattern again, known as "interference ripples." And if the water has been moving and then stopped this will lead to sediments forming a thin layer in the dips between the sand ripples; if this sediment is in any way different from the sand on the bed there, perhaps mud from an estuary, then it will create a color-rippling effect, known as "flaser ripples." Don't worry about being able to identify exactly what the water has done for every set of complex ripple patterns you see; just enjoy noticing them and seeing if you can draw one or two deductions.

AFTER A WAVE breaks and rushes up the beach as the swash, its energy finally peters out as a foaming, thin rush of water coming to a gradual halt on the beach, and at this point the sand that has hitched a lift will drop out. This sand will be ever so slightly different from the sand underneath, and this creates a visible record of the highest reach of the swash. These curved lines, which you can see on any sandy beach with waves, are known as "swash marks."

After the swash reaches its highest point some water sinks into the sand and some will begin a retreat back to the sea as the backwash. The backwash will create its own patterns, very different from the swash marks, typically more like thin diamonds that are about six inches long and stretched in the direction of the water's flow.

If there is a rising tide and the swash is especially big and strong it will surge up onto areas of dry sand. When this happens, some of the water will sink through the dry sand, and this forces the air in the gaps between the sand grains upward. This escaping air will burst through, creating a series of holes in the sand, known as "pin holes." If the air struggles to escape it will form bubbles under the top layer of wet sand and force them to bulge, creating a series of small domes. If you rest a finger on the domes they will collapse. You will see pin holes without domes quite often, but if you see the domes then there are very likely to be pin holes there, too.

Water that has soaked into the sand when the tide is high will begin to seep back out when the tide retreats. The water emerges from the beach face and runs down, and this creates another set of patterns known as "rill marks." These rill marks have certain telltale characteristics. Firstly, they spread out, starting as a main trickle and ending up as lots of thinner, much finer, little streams. This makes them appear to branch out, which can look like trees or, to my eyes, more like tree roots.

You may occasionally come across much larger features known as "beach cusps." These are crescent-shaped sweeps of sediment, from a few yards up to fifty yards wide. The crescents join together to form a series of cusps that touch at their horns. The coarser sediments collect at the horns and finer particles move into the middle of the bays. Scientists are not yet agreed on what causes them, which is both rather nice and a bit baffling for water readers. It is believed that the spacing of the cusps is determined by wave height, bigger waves leading to wider cusps.

It is well worth looking closely at the sand around any isolated rocks in the tidal zone, as you will get a chance to map the water's flow using the sand patterns there. I have seen some wonderful examples of reflection, refraction, and diffraction in the sand ripples that mimic the patterns happening in the deeper ocean around islands (see earlier chapter). In Cornwall, a county in southwestern England renowned for its relationship with the sea, there is even a traditional word for the calm patches of water in the lee of rocks, where there is shelter from the tide and waves: they call it the "spannel." After the tide has receded the sand under these spots looks very different from the other sides of the rocks. Even though the water is long gone, you can sense the earlier calm in the smoothness of the sand.

The best time to search for most of these micro beach features is early in the morning on a low and falling tide. You'll likely have the beach to yourself and the previous high tide will have ironed away the footprints and scuffs of the day before.

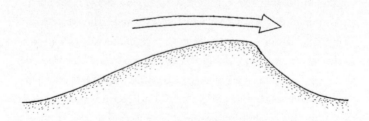

*Water has flowed from left to right.*

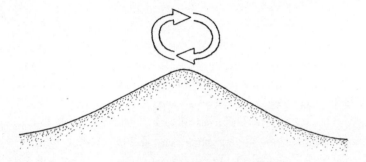

*Water has oscillated, probably formed under breaking waves.*

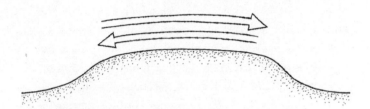

*Water has flowed one way with the tide, then the opposite way.*

## DRIFTING

When examining a beach at low tide, watch for the way the returning backwash isn't always a straight reversal of the swash. The direction that waves hit the shore will be influenced by a number of factors, including the bending of refraction that we saw in the last chapter, but the wind will usually be the decisive one. The backwash is not influenced by these things in quite the same way: It lives a much simpler life as it is pulled straight back down the beach by gravity. This means that depending on the relationship of a beach to the direction of the prevailing wind, you are likely to see trends of the swash pushing sand or shingle one way and the backwash not doing anything to reverse this. Over time this moves material along the beach consistently in the direction closest to the prevailing wind, and this is "long-shore drift" (so beloved of school geography teachers—it has to be up there with oxbow lakes as one of their favorites).

This transportation of sediment is a problem for those who think of beaches as being permanent and the coastal engineers that try to defend them. There is a strange truth in the profile of beaches: They have evolved in a physical sense to be a near ideal shape to defend themselves against the onslaught of the sea. This means that almost any attempt to engineer a "solution" to what nature is trying to achieve has as much chance of backfiring as working. Those groins that stretch down into the sea and are supposed to act as barriers to the beach being dragged one way are a good example. You never see just one groin, because a single groin will stop the beach rebuilding a little further along, and so it will exacerbate the problem there—so you need another one to compensate for a new problem created by the first. On a more positive note, you can at least look to see which side of the groin the sand or shingle is building up on, and that will instantly reveal the drift trend on that beach.

The combination of the swash pushing sand one way and the backwash pulling it back down another doesn't move all sand or shingle equally; it sorts it. Denser particles settle out of the water faster than the lighter grains, and so they don't travel as far on average, which leads to beaches with graded sand. As a simple rule of thumb, the darker the sand the heavier it is likely to be, as dark sands are normally composed of heavier minerals than the lighter-colored ones. It is therefore quite common for beaches to get slightly lighter in the direction of the drift.

Where you get a collection of beaches facing different ways—across an archipelago for instance—the effects of this sorting can be more dramatic. The Scilly Isles, at the southwest tip of England, are famous for their white sands and accompanying "tropical" light-blue waters. But the sands feel very different underfoot depending where you stand. In the west of the isles, at places like St. Agnes, the sand is gritty and makes a noise as you walk over it; further east, Tresco has powdery sand and Bryher's is more crystalline.

Studies have revealed that each patch of sand in the world is unique, which may go some way to explaining why we have a word for sand collectors—arenophiles—those who are gripped by its endless variety. (If you have felt the arenophilic pull you might want to check out The International Sand Collectors Society, motto: "Discovering the World, Grain by Grain.") We will never be able to pick up two handfuls with an identical mix of minerals and shells, and the water sorts these mixtures tirelessly, day and night until patterns emerge. The waves make maps of the sand for us.

One of the simplest types of sorting can be seen on those beaches where there is shingle down to near the water and then a strip of sand—these are the beaches that feel more "beachy" at low tide, when there is enough sand to lie on. The shingle is higher up than the sand because the waves are more powerful than the returning backwash. They push the bigger, heavier stones up the beach, where they come to rest, but the lighter sand gets carried back down again.

The action of the water on the sand is similar to the way miners pan for gold; the constant swirling leads to the heaviest particles—like gold—falling to the bottom and the lighter ones getting carried away. This is why experienced treasure hunters on the beach get down low at the start and end of the day and scan the beach, searching for places where the sand dips near to where the waves have been breaking earlier, as this will be where there's the most vertical movement of the sand and where the heavy items like gold will have collected.

The wind is also moving the sand, and again the lighter, smaller grains travel farther. The darkest, heaviest grains will drop out of the wind in any areas where the wind has a habit of slowing down, like in the lee of obstacles, so you get concentrations of dark sand in these spots.

Shingle behaves in a similar but different way from sand because the bigger stones gather momentum and travel farther than the smaller ones, leading to the stones being bigger at the end of the beach that the drift is moving toward. This sorts the pebbles so effectively that their size becomes a kind of map, pebbles the size of a fingernail at one end and bigger than your hand at the other end of long strips. Traditionally fishermen along Chesil Beach, in southern England, would work out where they were just by sizing up the pebbles, from tiny in the west to big in the east. I've read stories about Cornish fishermen who worked out where they were in fog from the "sing of the shore," the different sounds made by waves breaking on shores of a different nature.

## PLANTS AND ANIMALS

What do you think might be the strongest biological material ever tested? Spider's silk maybe?

> "People are always trying to find the next strongest thing, but spider silk has been the winner for quite a few years now," [Professor Asa Barber] told the BBC. "So we were quite happy that the limpet teeth exceeded that."

Limpet teeth are so strong that a piece of spaghetti made from the same material would be able to lift a Volkswagen Golf. Limpets, aquatic snails with conical shells and common in the UK, have a base, a home on a rock, which they leave at high tide to forage for algae and return to as the tide ebbs. Their habits are a bit more complex than just tide, though, as they vary their behavior according to both tides and daylight. These homes can be seen as "scars" in the rock when the limpets are away from home. Most coastal creatures have habits and rhythms that are tied to the tide. Oystercatchers, gulls, curlews, and crows know to scavenge along the beach on a falling tide, when there is plentiful fresh food in the sand.

Animals also give us clues to past events, both recent and more historical. Whelk cases, common on the beach in the UK, are whitish clumps of egg cases that look like a cross between bubble wrap and a sponge, and indeed sailors did once use them for washing, hence the nickname, "sea wash balls." Gray cases are a sign that the whelks have hatched; whelks might still be inside yellow cases. Whelks are cannibals, and the first to hatch will happily feast on their unhatched brethren. These cases are normally spotted during the January breeding time and are more common if there has been a storm recently.

A profusion of urchins, dogfish, and rays washed up on one section of shoreline is a twin sign: firstly and most obviously, there has been some rough weather. But the reason for the concentration of these animals along one beach could be a wreck just offshore. Wrecks form extraordinarily rich breeding grounds for these creatures and others.

Periwinkles, those small snail-like creatures you see on the beach, and especially common on New England shores in the US, are mapping the beach: the flat periwinkle is found low down in or near the sea's water, whereas edible periwinkles reside higher up as they need only an occasional covering of sea water, and at the top of the beach are the rough periwinkles that can survive long periods out of the water.

The creatures you find in rock pools will vary with each step toward the sea, too, because the height of the pool will determine not just how long it is fully under the sea's water, but also the concentration of salt due to evaporation. On a hot summer's day, the highest rock pools can have levels of salinity that would kill the creatures in the lower pools. The green crab tackles this problem by altering the salt balance in its own body.

Seaweeds also map the beach. *Silvetia*, bladder, and saw (or serrated) wracks are considerate seaweeds, especially the latter two, because they contain clues to their appearance in their name. Bladder wrack has bladders, and saw wrack has teeth. These three types of seaweed have each evolved to specialize in one band of the seashore. *Silvetia*, found on the Pacific, will be found highest up the beach, then bladder, and lowest is the saw seaweed. You can find the latter two along the Atlantic.

Red seaweeds have no chlorophyll and so are not dependent on sunlight, which allows them to survive in deeper water. However, they are often found washed up near the low tide level. One of the most common is dulse, which has been used as a food for centuries in parts of western Scotland and Ireland.

ON A ROCKY SHORE you will also spot the different environments as color-coded bands, where each color is a different lichen. At the lowest level, on the rocks that are underwater at high tide, you might find a black, tar-like lichen, called *Verrucaria*. Whenever an oil spill is reported, dozens of worried people report oil on the rocks—fortunately, most of these turn out to be the hardy black *Verrucaria* lichen.

Above this black band there are orange lichens, the *Xanthoria* and *Caloplaca* families. A little higher and the lichens turn gray; the crusty ones are *Lecanora* and the foliose ones, *Ramalina* and *Parmelia*. The easiest thing is to remember, "You get out of the sea, into a BOG"—Black, Orange, Gray. The more light, the more lichens, so this effect is most dramatic on south-facing rocky shores.

The strandline is the name given to the line of dead plants, animals, and flotsam that mix together and form the high mark on the sea's advance. If you think of sweeping the kitchen floor with a broad mop then the strandline is the curve of crumbs and dust at the leading edge. The strandline is usually a series of bowing curves because it is created by a series of strong swash marks.

The strandline is an unpopular spot for a beach towel as the density of decaying matter and the small creatures feasting on it make for strong smells and yelps, but they are well worth investigating. In among the other flotsam, you will find some driftwood, and a rough gauge of its time at sea can be made from the way it has been smoothed and the number of worms or barnacles that have made their home there. (At the time of writing, there has just been an extraordinary coincidence. This method of dating flotsam from the barnacles on it has just featured across the world's TV stations and newspapers. The first piece of debris from what is believed to be Malaysian Airlines flight MH370 has just washed up on Reunion Island. Experts are suggesting that the part of the wing that was found, the "flaperon," has the right number and type of barnacles on it to be consistent with having been at sea for over a year, long enough to be from the missing flight.)

In the Pacific, David Lewis encountered some extraordinary weather lore, where the Islanders would use it as an aid to planning the time to set out on their voyages. One *tia borau*, or navigator, on the island of Nikunua in the Gilbert Islands, explained in detail how the behavior of the crabs was used as a weather forecast. A crab that blocked the mouth of his hole and scratched the sand down flat over the opening, creating marks like the sun's rays, meant that wind and rain would arrive within two days. But if the crab made a pile of the sand, without covering the hole, then there would be wind but no rain. And if it blocked the hole, but didn't flatten the mound of sand or leave scratch marks, then there would be rain, but no wind. Only if the crab leaves the pile of excavated sand and his hole unmolested will the weather be fine.

The beach ants forecasted the weather for the Islanders in the way they treated their food. Food left in the open indicated good weather to come, but food hidden in homes or shelter of any kind meant deteriorating weather in the coming days. There were additional clues in the places that spiders spun their webs and the behavior of the starfish on the reefs. Even the reefs themselves were said to foretell of changing weather, emitting a clear fluid before clear weather and a dark or milky one if the waves were going to be big.

WE WILL LEAVE the beach after trying and failing to dry off after a swim, thanks to a phenomenon known as hygroscopy. Have you ever noticed how on a beach you never feel 100 percent dry again after you've been for a swim in the sea—there is always a dampness, a clamminess, long after the sun should have dried you? Hygroscopy is the name given to the way certain substances attract water. Salt is a hygroscopic substance, which is why salt cellars often have rice grains in them or other ways of keeping the moisture away from the salt. It is also why a swim in the sea keeps us moist for a long time—no sooner have we dried in the sun, but the salt on our bodies starts attracting the moisture in the air onto our skin again. Something to contemplate alongside that other, better-known, but less scientifically intriguing beach phenomenon of there always being sand in our picnic sandwiches.

# 15

# Currents and Tides

A T MIDDAY ON SEPTEMBER 19, 2010, a small group of spectators gathered at Trinity Buoy Wharf by the River Thames to watch and listen as a bell was due to ring for the first time. The new bell is called the Time and Tide Bell, and it was designed to ring as the waters of the river washed over the base of the bell at high tide. The crowd that day were disappointed, the bell stayed silent, the water had failed to reach the bell.

"What has rather complicated today's event is that there's a big high pressure . . . that was something I never knew about . . . the amount high pressure affects the tide level . . .," Marcus Vergette, the Bell's sculptor, told the bemused audience, ". . . it should actually be ringing right now."

IF YOU LOOK at a sea wall for long enough you will notice water flowing past it and the sea-level change. Most people quickly work out that these two effects must be due to the currents and tides. But there is a mischievousness about the expression "currents and tides"; it is used collectively by many to paper over a multitude of ignorance.

Currents refer to the flow of water in a horizontal direction. Tides refer to the cyclical change in the height of water, driven by astronomical forces, like the moon. Hold onto those two basic definitions because they can get you out of a mess.

Understanding what currents and tides are in general terms is quite easy, but working out exactly how one patch of water will be influenced by them is a different challenge altogether. It is a challenge that the water reader cannot shy away from and in this chapter we will move step by step from the basics to a level that is not rocket science, but that very few people on Earth ever attempt. To demonstrate this point, there is a game you can play with a sailor. Just ask them the following questions, then watch their faces contort, their answers get tied in knots, before they turn away, then run and jump into the sea:

> Why, if the moon only goes round us once in twenty-four hours, do we get two high and low tides each day? Why is the second high tide each day sometimes a noticeably different height from the first?

## CURRENTS

In May 1990 a storm washed containers off a freighter, leading to the loss of 61,820 Nike sneakers. Over the following months these shoes started appearing on beaches, giving oceanographers an extraordinary insight and rare opportunity to map their journey. A couple of years later 28,800 rubber ducks (well, rubber ducks and floating toys anyway), were washed off a ship and began *their* bid for freedom. Tiring of a life at sea, the rubber ducks began pitching up on beaches ten months after the spill. They were found from Hawaii to Iceland; one suspect arrived in Scotland eleven years later. Thanks to sneakers and toys we know a lot more about ocean currents than we did a few decades ago.

Currents will flow in water whenever there is an imbalance. When the sun heats the sea it creates two imbalances, in temperature and salinity, and this leads to water in certain areas becoming denser than in others. The Mediterranean Sea is one of the best demonstrations of this effect. The sun causes the confined sea's water to warm and evaporate more quickly than the Atlantic, making the Mediterranean sea-level sink, but also making it saltier and denser than the water in the Atlantic. This sets up two currents, one near the surface where Atlantic water flows in past the Gibraltar Strait to "refill" the Med; the other, much deeper current takes dense, salty water out into the Atlantic. These ocean currents are known as "thermohaline" and are comparable to the way the sun heats the atmosphere, changing temperature and air pressures, which is what generates the winds.

The major ocean currents are not often seen, but the principle is still important: if there is any change in temperature, salinity, or density, this will influence the way currents behave. The most likely place you will witness this is by the coast, especially where estuarine water meets sea water. If you can spot the color change that marks the boundary between these two different water types, then look closely and you may also find evidence of the currents behaving differently, too.

THE MAIN CAUSE of ocean currents, and the one we are most likely to be able to read the effects of, is the wind. We have seen how wind will create waves, but it also sets up currents. When you blow into your tea, long after the ripples have died away there is still some movement in the tea—you'll spot it most easily if you do this just after adding some milk, but before stirring it; the milk "clouds" will help you to spot the swirling movement of the currents.

Water at the surface gets pushed by the wind but the "stickiness" of water molecules that we looked at near the start of the book means that this surface water pulls its comrades along with it. So it is not just the very top layer of water that starts moving, but a thicker layer, down to about 100 yards in the oceans. The longer the wind blows, the stronger the wind, the shallower and warmer the water, the faster the current will be. Over a deep, open ocean it is typical for wind-driven currents to struggle to get much above 2 percent of the wind speed, but in shallower, warmer water it will have a more noticeable effect. A 10-knot wind blowing over warm water that is only 1 yard deep will be able to create a 1-knot current, 10 percent of the wind's speed. Since most currents are in deep water, the global average is only about half a knot. Wherever there is a wind-blown water current, it will be strongest nearest the surface and grow progressively weaker with depth.

If you look at a map of the ocean's currents you will notice that they nearly always follow a curved path and that there is a clockwise tendency in the northern hemisphere and a counterclockwise one in the southern hemisphere. This is because anything traveling significant distances over a rotating sphere, in this case Earth, will be deflected, a tendency called the Coriolis Effect. For long distance oceanic currents this can mean a deflection of 45 degrees from the direction of the prevailing wind.

It is time to deal with one of the little nautical foibles that might otherwise trip you up if you are unfamiliar with it. The convention is for wind directions to refer to the direction they have *come from*, but for water currents to be described in terms of the direction they are *heading toward*. A westerly wind therefore creates an easterly current.

There is one big challenge for the water reader when it comes to currents: They are often almost invisible. Unless you are lucky and the current is carrying water of a very different character into a new home, as in the case of estuaries like the Nile pouring silt into the Mediterranean, or the striking dark blues of the Kuroshio Current and Gulf Stream, then water currents are hard to detect. Hard, but not impossible. If water is doing anything then there will be a way to read this. It is just that with currents, this is a very fine art. For example, let's imagine two similar but different scenarios. A current is moving water at 2 knots on a still air day. This will have the same effect on the surface of the water as if the water were stationary and a gentle breeze of 2 knots was blowing over it: Tiny ripples would be created, and any ripples when there is no breeze must mean the water is moving.

Now imagine there is a 2-knot breeze blowing over the ocean and this has created some ripples. If there is a narrow current moving in the same direction as this wind, it will create a calm stream, without any ripples, because there is no apparent wind over that water—it is moving at the same speed as the breeze.

There is no denying that it takes beady eyes to spot either of these effects, but they are out there almost every time we look out to sea.

Currents will have a small effect on the wavelength and height of any waves that have formed. A current flowing in the same direction as the waves will stretch the waves and flatten them slightly, and an opposing current will bunch the waves up a little, compressing the wavelength and raising their height. Strong winds against strong currents create dangerously choppy seas, but most of the time the effect is subtle.

The contemporary Hawaiian navigator Nainoa Thompson, of the Polynesian Voyaging Society, reported noticing a change in the water's behavior when on the open ocean. He had to depend on his experience to decide if this was due to a change in the wind or whether the current was now running against the wind, effectively going against the grain and creating a slightly rougher sea. Expert racing sailors will scour the water to see if they can tell from the shapes of the waves where the strongest and weakest currents might be and then take this into account in their race strategy.

One of the best places to practice looking for this effect is actually on a river. On days when the water is flowing quickly one way, but the wind is blowing the opposite way, take a good look at the shape of the ripples on the river. You will notice that they have their own character, I think of it as being a bit more "jagged" and harsher. Once you get to know it, you will recognize it without needing to feel or think about the wind, and it is very satisfying being able to spot and instantly identify a wind against current shape to the water's surface.

The ocean's currents can carry anything at the surface, like the sneakers and rubber ducks above, on long journeys. But they do not carry all things equally. Dutch oceanographers discovered that Wellington boots that were lost by fishermen in the North Sea did not all embark on the same journey. The currents carried the left boots to the east, washing them up on the Dutch coasts, but the right boots were taken west to Scotland. The shape of flotsam determines how and therefore where it will be carried.

On the other side of the world, on the Ka'ū coast of Hawaii, there was once a strange and macabre tradition for those who had lost loved ones at sea. The locals would search two different stretches of beach, depending on the social status of the person who had drowned. This was not part of some religious or superstitious practice, but because the rich and poor genuinely did wash up on different beaches. The line of beach known as *Ka-Milo-Pae-Ali'i*, which translates roughly as "the twisting water washes ashore royalty," was the favorite for upper-class corpses, and one further around, *Ka-Milo-Pae-Kanaka*, "the twisting water washes commoners ashore," was the preferred resting place of commoners. The currents were sorting the fat, rich bodies from the poor, lean ones.

## TIDAL CURRENTS

Sir James Lighthill was one of those brilliant mathematicians who improved our understanding of a specific area significantly. His field was fluid dynamics, and he did pioneering work that helped decipher how waves behaved in areas as diverse as mud flows and traffic jams. But Lighthill's interest in the way things flowed was not limited to dry academia. He studied the way the water flowed vigorously around the Channel Islands, the archipelago of islands in the Enligsh Channel, and then put his observations to the test. In 1973 he became one of the first people to successfully swim the eighteen miles around the Channel Island of Sark. This formidable challenge has surprisingly little to do with the distance swum and a great deal to do with being able to successfully predict how the water will flow—it moves much too quickly for anyone to be able to successfully swim against it for any time. Sir James repeated his Sark circumnavigations several times successfully, until his heart gave out on his sixth attempt and he died wrestling with the tidal currents in July 1998, aged seventy-four.

What exactly does the expression "tidal currents" mean? If we think back to the two definitions at the start of the chapter, the answer will be found by combining them. Tidal is just the adjective for tides, of course, so tidal currents refer to the horizontal flow of water resulting from the change in height of water caused by the tides.

If water is higher in one place than another, then gravity will make it flow toward the lower height. Since tides make the water high in some places relative to others nearby, there is a flow of water toward these lower areas. And if the conditions are right, this flow can be extremely powerful indeed. The strongest in the world is in the Saltstraumen, near the town of Bodo, Norway, where millions of tons of seawater surges through the strait at up to 22 knots.

In any coastal areas, tidal currents are normally the strongest currents, dwarfing any effects from density, salinity, temperature, and even wind. These are the currents that coastal sailors have to be most wary of and the reason why any decent chart of coastal areas will include information about these currents.

Tidal currents are such an important aspect of coastal seafaring that they have spawned their own subset of the nautical language. There are tidal "races," where fast tidal currents are made even more worrisome by being forced through narrow constrictions, leading to turbulent seas. And tide "gates," where there are windows of opportunity for boats to pass through, but that effectively shut at other times because the flow of water is too strong and unfavorable. I once had to skipper a boat from St. Malo in Brittany, France, which you can only leave at certain stages of the tide, past an area with tidal whirlpools, to St. Helier, on the largest Channel Island, Jersey, where you can only arrive at certain stages of the tide. It takes a lot of table-consulting, scribbling, and head-scratching before setting off in such places, if you want to avoid becoming a toy for the tidal currents.

One of the easiest indirect ways of detecting tidal currents is by looking at boats at anchor. Boats will be swung around their anchor or mooring by both the currents and the wind, but in tidal areas these currents are normally the most powerful of the two, and boats can be seen pointing dependably into the current like a weather vane. There are few things more pleasurable than looking out over a harbor at the turn of the tide and watching the boats swing slowly around on their moorings to point the opposite way—for the water reader this moment is the cuckoo coming out of the clock.

You can also look for the difference between the direction a slow boat is pointing and the direction it is actually traveling.

If a boat is sailing at 5 knots through a crosscurrent of 2 knots, it will need to head 20 degrees into the current to continue in the direction they want to travel (sailors have to make these calculations before setting off, plotting something known as their "course to steer"). This is something that is even easier to see from on board, but can be spotted from shore, too. You will see the same effect in a very different place if you look up at aircraft landing on a windy day; if there is a crosswind, the pilot has to point the aircraft in a different direction to the runway until the very last moment.

Tidal currents do not flow at an even speed—they don't flow one way at a certain speed, then flip and flow in the opposite direction at the same speed. What happens is that they are almost constantly accelerating and then decelerating. From a standstill, known as "slack water," at both high and low tide, the flow accelerates steadily until the period halfway between high and low tide, when it will be moving its fastest, at the peak flow rate, but as soon as it reaches this speed it begins to decelerate again, all the way back to another slack water. Put simply, the closer you get to high or low tide, the slower the water flow will be, and during the two hours in the middle the water flows at its fastest. The flow of water during spring tides is usually close to double that during neap tides (spring and neap tides are explained shortly).

The rate of change of tidal currents catches a lot of people unawares. One of the classic and dangerous situations that arises is when one group of swimmers decides to copy another group, assuming conditions will be identical. I remember a time in the Solent, the strait between mainland England and the

Isle of Wight, when one group of swimmers enjoyed a leisurely swim and the second group followed an identical route, setting off only ten minutes later only to find themselves struggling. I remember it particularly clearly, because I was in that second group. At one point we found it hard to swim forward against the current at all, even though our friends had just moved through it with apparent ease. Ten minutes can be enough to make the difference between swimmable and dangerous.

IN PARTS OF THE WORLD where the sea is an integral part of life and there are few other dependable references, the tidal currents are befriended. From the Warao people of the Orinoco Delta in Venezuela to colder climes, the flow of the water is seen as an aid to navigation, not a threat to it. The Warao view the world in terms of upriver or downriver, toward the sea or away from it, and use their sensitivity to the direction the river is flowing as a vital aid, helping them to orientate themselves.

In the Arctic, the Iglulingmiut, an Inuit group, detect which way the currents are flowing by looking at the *qiqquaq* fronds (kelp), and then using their intimate knowledge of the water's habits to find their way from this alone. They are experienced enough to be able to tell the difference between the main current and the back eddies that are formed close to the shore, which would otherwise send them in the opposite direction from the one they needed.

## THE TIDES

Coastal water goes up and down regularly, and it's got a lot to do with the moon. I think that is a fair summary of popular understanding of the tides. Actually I'd go further; that is a summary of even many experienced seafarers' understanding of the tides. Many sailors learn how to predict and calculate tidal heights at ports with some accuracy, but very few are either taught or take the time to understand the cause of those heights.

From 1833, when the first tide tables were produced by the Admiralty, the emphasis shifted from looking, thinking, and understanding, to depending on tables of others' measurements.

WE GET TWO high and two low tides in any one place in any twenty-four-hour period, as highs replace lows and vice versa. We find high tides and low tides on opposite sides of the earth at the same time. Tides can be thought of as a pair of very long waves that ride around the earth. These waves straddle half the globe between crests, have a height of only a couple of feet, and travel at between 700 and 800 miles per hour. But why do these waves exist in the first place?

The moon is a lump of rock that appears over roughly the same part of Earth once every twenty-four hours and fifty minutes. It is fairly small relative to Earth, with a volume about the same as the Pacific Ocean, but it is very close, and so it exerts a significant gravitational pull on us. Most things stay firmly rooted to the ground, fixed there by Earth's own, much stronger gravity, but large areas of water are more fluid and react to the moon's pull.

The large bodies of water directly under the moon are being gently pulled away from Earth by the moon, which creates a little bulge in that sea. This is the high tide that makes sense to most people. But why is there a high tide on the opposite side of the planet—if the high tide is caused by the moon attracting the water, then that is very odd, surely? The answer is that the moon is not just pulling the water, but it is pulling everything on Earth, including Earth itself and all the water on it, even the water on the opposite side. But—and this is the key—the moon is so close to Earth and Earth so big relative to the moon that it means the moon's gravitational strength is much stronger on the side of Earth nearest to it than the far side. The moon pulls the nearside ocean very strongly toward it, creating a bulge and a high tide, it pulls Earth a bit less, so Earth gets left behind a bit and then it pulls the far-side ocean even less, leaving it further behind. It is the way the opposite ocean gets left behind that creates the high tide on the opposite side of the world. As Earth spins on its axis, these bulges rotate around the world, giving us two high and two low tides every twenty-four hours.

(Strictly speaking the moon doesn't orbit Earth, the two are orbiting each other around a shared center of gravity that is 2,900 miles from the center of Earth. The writer James Greig McCully gives a lovely analogy of this: Imagine a girl on ice skates with a long ponytail spinning around very fast with a bucket of water held in outstretched arms. The water would be pulled outwards, sticking to the bottom of the bucket—one high tide. Her ponytail would also be pulled outwards by the centrifugal force, high tide number two.)

It may seem weird the first time you encounter this expla-nation, but bear with it, because so few people understand this that you are now entering a select and secretive cabal, the small group of people who actually know what a tide is. Alexander the Great was mystified by the tides and even Galileo misunderstood them, so we should all forgive ourselves if we find them challenging at times.

Since the moon is dictating the high and low tides, we should expect there to be a high tide whenever the moon is closest to overhead. On the open ocean this is largely true, there is a very small delay of only about three minutes, because the water is not being impeded by the land. However, in coastal areas there is much more friction and so the bulge gets delayed significantly. In practice what this means is that there is always a dependable relationship between the moon passing its highest point and the high tide in that location, but the time difference will be specific to that place. All you need to do is notice how long after the moon is highest in the sky that the high tide arrives and you have a yardstick for that place forever. It may be minutes or hours, but it will be fairly dependable.

The moon rises fifty minutes later each day on average, because it orbits Earth slowly from west to east and so slips back a little relative to the sun. Put another way, if you were to watch the moon and sun pass over a church steeple and you started a stopwatch, it would take twenty-four hours for the sun to pass the steeple again, but twenty-four hours and fifty minutes, on average, before the moon passed over that steeple again. The moon has a cycle that lasts 29.5 days, which is the time it takes to get back to the same place relative to the sun. These two periods dictate two of the main tidal rhythms: high tides arriving fifty minutes later each day on average and the full cycle lasting approximately one month. (If it helps, think of the sun and moon starting a race together at the time we call a new moon. But the sun is always faster and edges ahead 12 degrees or one-thirtieth of the circle every twenty-four hours, until it laps the moon after a month and the cycle starts again.)

THE EFFECT OF the moon and its monthly orbit give us the basic template of the tides, the pattern of the highs and lows over a month. Next we need to understand why tidal heights can vary so much over the course of a month in the same location. After the moon, the next most important influence on the tidal cycle is the sun. The sun is 27 million times more massive than the moon, so we might expect it to have a much greater effect, but it is so far away, 400 times further from us than the moon, that its gravitational effect is diluted to the point where it becomes a more minor, but not insignificant, player.

The sun exerts half the gravitational effect of the moon on our oceans. In the center of a big ocean, the moon is able to pull the water up into a bulge, perhaps a foot high, and the sun is able to pull the water up half that, about six inches. If the sun and moon are both pulling Earth and its water in the same line, then these forces are joined and the ocean bulge will be the two bulges combined, eighteen inches. But if they are pulling across each other, the effects are minimized. The sun and moon are in line twice a month, at a new moon and a full moon, and shortly after these we experience the greatest tidal highs and lows, and these are known as *spring tides*. When the sun and moon are not reinforcing each other's pull, we get the lowest high tides and the highest low tides—in other words, the smallest range. These are known as *neap tides*.

I'll admit that I found it hard to fully understand the logic behind spring and neap tides for years; I found it very odd that neap tides happen when the sun is pulling in a direction that is 90 degrees from the moon. It seemed counterintuitive to me, as I thought it would make more sense if neap tides were when the sun and moon were opposite each other. The discovery that clarified this for me and which I hope will help you, too, is that when the sun is 90 degrees from the moon, instead of two very high and two very low tides circling Earth, what we have in effect is two fairly strong high tides, caused by the moon, and in between them two weak high tides caused by the sun. There is only so much water to go around and the moon pulls most of it, so the sun's high tides end up as what we call neap low tides. Don't worry if that doesn't help you, but it's just one more perspective on a fairly complex area.

We must cut ourselves some slack when learning about the tides; it's a tricky area. In the Second World War Battle of Tarawa in the Pacific, the US Marines faced a daunting battle against determined Japanese defenders, and it was made even tougher by the fact that "the ocean just sat there." Planners had failed to appreciate how a neap tide would mean there was less water over the reef than they had expected. This was compounded by the fact that the moon was very far from Earth, too, further weakening the high tide. Many of the boats were unable to clear the reef and were stopped short, leaving the attack exposed to bombardment. Many more lives were lost than was necessary, and there are still hunks of jagged metal in the reef to this day.

Much more recently and well away from the fog of war, a car carrier, a large type of cargo ship, was beached on a sandbank in the Solent, near Southampton, in January 2015. The *Guardian* reported what happened next:

> Initially, the idea on Wednesday had been to deliberately refloat the ship because it was feared it could sustain further damage if it remained on the sandbank, where it was resting lightly and was likely to be battered by the bad weather currently hitting the UK.
> This scheme was abandoned because it was felt there was not enough time to clear water from the ship and instead the plan was to anchor the vessel down and refloat it on a future date. But the high tide and gusting winds caused the vessel to refloat itself. Officials said the boat was out of control briefly until it was stabilised and towed to safety.

Here we can see how some the world's best marine salvage experts drew on centuries of experience, pored over data about the weather, and looked at computer predictions about tidal heights, and then even they couldn't tell when the tide would lift a boat off the sand.

The good news is that most of the key moments in the moon's cycle are easy to spot. A full moon means you are not far from spring tides, the highs will be very high and the lows very low. If you see half a moon, either side bright, then you are not far from neap tides, and the highs and lows will be less dramatic.

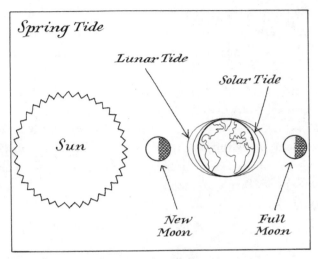

*The earth, moon, and sun are aligned creating "spring" tides.*

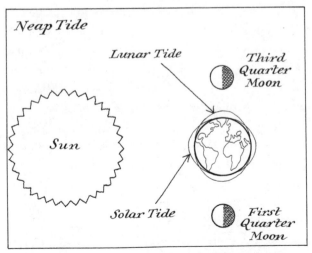

*The earth, moon, and sun when they are creating "neap" tides.*

A new moon is when the sun and moon are roughly aligned, and at this time the moon cannot be seen with the naked eye, but if you know that you are near a new moon then you are near spring tides.

The distance of Earth from the moon and sun varies, as the orbits are elliptical, and the closer we are to them, the stronger the gravitational effect. The sun and moon are also over different latitudes at different times, so on top of the fortnightly swing from spring tides to neap tides there are both longer and shorter cycles that act on top of these rhythms to make each tide more or less extreme.

This is why you will come across expressions like:

- Equinoctial tides—these extreme tides happen when the sun is over the equator. Expect unusually high and low spring tides toward the end of March and September.
- Perigean spring tide—another extreme, when a spring tide coincides with the moon being at its closest point to Earth.
- Supertides—an 18.6-year cycle of the moon's position relative to Earth that leads to tidal extremes. 2015 was a year of supertides. The next will be 2033.

On February 5, 2004, twenty-one Chinese laborers drowned while picking cockles—small clams—off Morecambe Bay in northwest England. Tragically these workers had been sent out into an area that fluctuated between being land and sea and had been left to work there too long.

If you look at any chart of a coastline in a tidal area you will discover a zone that sits uneasily between the land and sea. On the chart, this will be a different color from the land and the sea, and this third color signifies an area known as a "drying height." Drying heights are the areas of land that are dry at times and under seawater at others. The fluctuation in the height of tidal water creates these nether regions, and if the coastal gradient is shallow and the tidal range is high then they become quite big in places and can stretch for hundreds of yards.

These areas will generally all be sea during a high spring tide and land during a low spring tide, but in between there is a constant wrestle going on between the two. The ambiguity between land and sea has opened the way for some creative uses of the temporary space; there have been "Reclaim the Beach" parties along the Thames in years gone by, revelers enjoying a narrow strip of wet land for a celebration, sandwiched between prime waterfront real estate and the Thames. And in the Solent, there is a sandbank called the Bramble Bank that is uncovered at low spring tides. It was on the Bramble Bank that the car carrier that we met earlier ran aground. This small temporary island has traditionally played host to a sporadic cricket match between two sailing clubs, but the full match is rarely completed as the tides are not a great respecter of cricket.

At low spring tide, parts of a beach that are only exposed for a few hours each month are worth a visit, and one of the extreme springs, equinoctial or perigean, can uncover land that is underwater for months at a time. If you are lucky you may witness shipwrecks or petrified forests emerging from the sea.

IN THIRTY YEARS of passionate interest in the seas I can't recall ever coming across someone who was familiar with the reason behind the following, very common, phenomenon. On many days the two high tides we see will be of noticeably different heights. Obviously this can't have anything to do with the position of the moon relative to the sun—which explains springs and neaps—and it can't have anything to do with the distance of the moon or sun from Earth, as neither of these change markedly over twelve hours.

The answer lies in something called "lunar declination," which fortunately is much simpler than it sounds. The moon only ever appears over a section of the Earth's surface that stretches either side of the equator. This section varies in width, but at its maximum it can be thought of as being slightly bigger than the tropics (in latitude terms it is 28.5 degrees north and south of the equator). Put another way, you will never see the moon overhead if you are north or south of Africa.

When the moon is over the midpoint of this range, it is over the equator and the two high and low tides will be close to equal, but when the moon is near its northerly or southerly limits, then things become a bit imbalanced, and one tide will be noticeably higher than the other. The only way to tell this by looking is to glance at the moon as it passes south of you; if it appears abnormally high or low in the sky for that aspect, then the chances are that the moon is well north or south of the equator and the tides will be uneven.

Incidentally, the time it takes from one maximum range of lunar declination to the next is 18.6 years, which is where we get the timing interval between the "supertides" above.

The moon and sun together create the long, low waves that roll around the globe setting the rhythm for the high and low tides, but they cannot explain why we get such huge variations in height and behavior from place to place. The solar and lunar rhythms are so dependable, regular, and predictable, and yet the tides we see at the coasts around the world are so varied that it's hard to imagine they are connected. There are two main considerations that help to explain this variability: the size of the sea that the coastline is neighboring and the shape of that coastline.

The larger the sea the more water will be in the tidal bulge, so small seas cannot generate huge tidal ranges. There is no mention of tides in the Bible as the Mediterranean is too small to generate big tides. But on the other side of the Arabian peninsula, we find the following references to the sea's tidal behavior in an account by a first-century trader:

> Now the whole country of India has very many rivers,
> and very great ebb and flow of the tides; increasing at
> the new moon, and at the full moon for three days, and
> falling off during the intervening days of the moon. But
> about Barygaza it is much greater, so that the bottom
> is suddenly seen, and now parts of the dry land are sea,
> and now it is dry where ships were sailing just before;
> and the rivers, under the inrush of the flood tide, when
> the whole force of the sea is directed against them,
> are driven upward more strongly against their natural
> current, for many stadia.

One of the most decisive factors in the height of tide that you see at the coast will be the local topography. As the tidal bulge comes into contact with the land, all sorts of wonderful things start to happen. The water rushes up river inlets, slops up against coastlines, swirls around islands. It even sometimes bounces back and cancels itself out, which is why there is only one high tide in some places, like parts of the Gulf of Mexico. In Southampton Water, the estuary on England's southern coast, there is a phenomenon known as a "double tide," where the topography and resulting flow of water leads to a double peak at high tide, effectively an extra-long period of high water— one of the things that helped Southampton grow as a naval and commercial port.

The sun and moon together can only lift the tide by eighteen inches; anything more dramatic than that must be explained by the shape of the land and the way the high-tide bulge reacts to that shape. In a few cases, the land can concentrate this small bulge into the extreme heights of fifteen yards or more that we see in places like the Severn Estuary, in South West England. The reason that the Severn Estuary sees such wild ranges becomes very clear if you look at it on a map: It's a narrow funnel, and whenever water is forced from wide-open, deep areas into tight, constricted shallow ones, then speed, height, and behavior of the water all get wildly exaggerated.

There are a couple of basic rules for working out whether the differences you see in tide behavior are due to the sun and moon or due to local factors. If you see the height and pattern of tides doing different things on different days at the same location, the reason for that will be found in what the moon and sun are doing. If you see the tides doing different things *on the same day* in different locations, the reason for that will be found in the shape of the coastline.

Tide heights do not move from low to high water at a constant rate, and their rate of change will follow a simple pattern, which can be remembered with the help of a very useful trick called "The Rule of Twelfths." There are approximately six hours between each low and high tide, and the total amount of water that arrives in that time can be thought of as being split into twelve equal volumes that are shared out among the six hours unevenly. In each hour after low water, the following amount of water arrives:

$$^1/_{12}, \; ^2/_{12}, \; ^3/_{12}, \; ^3/_{12}, \; ^2/_{12}, \; ^1/_{12}$$

As you can see from these fractions, half the water arrives in the two hours in the middle ($^3/_{12} + ^3/_{12} = ^6/_{12}$) and the height changes very slowly as you get near either high or low tide. This is the same effect that we saw with tidal currents, only here we are referring to vertical changes, not horizontal—heights not flows.

At each high or low tide there is a momentary calm, the pause before the water reverses and begins to accelerate up or down again. It is known as slack water or the stand of the tide in the West, but all seafaring cultures have their own view of this stasis. Among the First Nations people of British Columbia, this time was called *xtlúnexam*, and its calm was used within stories to signify that all would be well in the end.

THE RELATIONSHIP BETWEEN the earth, moon, and sun dictates the tidal rhythms that we see, and the shape of the coastline explains most of the variability from place to place on the same day. But there are a host of other minor factors that can have a bearing on the height of the tide. Individually they don't often amount to a big difference, but if they happen to coincide, then together their effects can be compounded.

The wind will pile water up on lee shorelines, and in extreme cases this can create something called a storm surge, when tides shoot over their expected heights and can cause huge damage to life and property. In 1953 there was a storm surge that raised sea levels by several meters in the North Sea, leading to more than 2,000 people losing their lives along the coasts of Holland and Britain. And in the US, the winds of hurricanes regularly lead to enormous storm surges. Hurricane Katrina, in 2005, produced a surge twenty-eight feet high as it pounded New Orleans, and left a death toll estimated as high as around 1,800 people.

Strong winds are usually generated by areas of low air pressure, which can then compound the problem. As we saw earlier, the air pressure will influence the height of tide: The lower the pressure, the higher the tide will rise on any given day. You may find it easier to remember this by thinking of high pressures sitting on and squashing the water and lows letting that water rise up, like squeezing one end of a long, thin balloon. This is normally a minor influence, but an extreme low can add a foot to the tide at the coast.

The winds will affect the tides well beyond the area where they are currently blowing. For this reason, if the tide is higher than expected or predicted and low pressure doesn't fully explain this, then this may be a warning of strong winds out to sea and of bad weather approaching.

Added to the wind and pressure, the temperature of the water will affect the tide. If the water is very warm this can add six inches to coastal levels.

The wind can also have an impact on the time the tide turns—a wind blowing in the same direction as the tidal currents will delay the turn and one against it will accelerate it, in each case by up to an hour.

On coastal rivers, the farther from the sea you go, the later the tide will turn on average. It may turn ten minutes later, if you are a few miles inshore, than it did at the coast, or even by an hour if you are thirty miles inshore. This can have the bizarre consequence of a river temporarily running in different directions *at the same time* in locations that are only a few miles apart.

THERE ARE MANY asymmetries in the behavior of tidal rivers. The ebb is usually stronger than the flood as the rising tide battles with the natural downhill flow of the fresh water. When the ebb finally prevails there is a backlog of built-up water that then surges out toward the sea. And as with all rivers the flow is uneven, much faster in the deep areas and slower at the shallow edges.

If you are looking at a tidal river, there are a couple of tricks worth knowing. These rivers deposit twigs and all sorts of other flotsam at their high-water marks, particularly in any convenient traps by the water's edge, like a fork in a willow's branches. But being beady-eyed, you will hopefully start to notice how the amount of flotsam you can see in the water—twigs, leaves, rushes, and detritus—shoots up when you are near spring tides. During spring tides the water reaches higher than it does in the intervening fortnight, which means that once a fortnight the water sweeps up a lot of random matter from along the riverbanks. This ends up in the fastest part of the river's stream, the center, for a while, before it then ducks out of this in an eddy or sheltered patch, where it gathers. So you find more flotsam both in the water and gathering in heaps at the edges near spring tides.

If you get to know a tidal river well enough, you will quickly notice how the level of detritus, natural and human, in these spots fluctuates, as the height of the water and its speed vary over time. It is strange to say, but you can actually estimate the moon's phase by looking at the amount of flotsam on any tidal river. The greatest tidal ranges and speeds come after new and full moons, so if you see a lot of flotsam on the river and gathering in these dead-water patches, then you know you are close to a full or new moon. After a spell of very good weather recently, I spied a deckchair floating serenely, but quickly, down the middle of the river; it was not a coincidence that it had been a full moon the night before.

You will know the pieces are starting to come together for you when you spot these small signs and they send your mind off on a deductive spree. "Ah! Lots of twigs on the river, it must be spring tides, which means that we are near either a full or new moon." When you see the full moon later that evening, it's not a bad feeling at all.

## ANIMALS

So many animals are tuned to the tides that you may be able to spot a distinct change in animal behavior as the tide turns. A friend reports that the cormorants are much more common on the Thames during the ebb tide than the flood and that when the tide turns at low tide, the cormorants fly away. On a related note, I have come across a few reports of people being able to hear the tide turn, but I haven't had that pleasure yet and it remains on my wish list. The closest I've managed is the bubbling in mud flats as gases escape at low tide, but you may do better.

Many animals are tuned to the powerful effect of tidal currents. Claudius Aelianus, a Roman author who wrote about the behavior of animals in the third century, noted how crabs would not try to battle the currents that accelerate around headlands, because:

> . . . they are aware of these dangers beforehand, and
> whenever they come close to the headland, each one
> heads for some sheltered recess and waits for the others.
> Once they have assembled in this spot, they crawl up
> and onto the land and scramble over the cliffs, and
> thus by land they pass that point where the sea has the
> strongest currents and is most forceful.

I particularly enjoy this ancient example as it chimes with the behavior of otters we saw in the Rivers and Streams chapter, the way the otters will swim with the flow of the water when heading downstream, but take shortcuts over land when heading upstream.

Since all coastal creatures' lives are influenced by the tides to some extent, the best approach is to focus on the few that you find most often or that you find most fascinating. It is not possible for me to list all of the intriguing relationships that you may discover, but I can offer one example I've grown quite fond of, the lugworm, that illustrates the idea. They're common in the UK, but even if they're new to you, they're easy to visualize: The sand "castings" they leave behind all over a wet, sandy beach look a little like a miniature, sand-colored cross between a dog turd and a pile of soft serve ice cream.

The two main lugworm species are bunched together, which is common with all animals and plants when we first meet them (did you know there are 375 species of blackberry?). But specialists learn to differentiate between two main ones. Lugworms make good bait and anglers tend to know best, they name the two types as Black Lug and Blow Lug. Black Lug have a neat, round cast, coiled in an orderly way. Blow Lug leave a more anarchic mess.

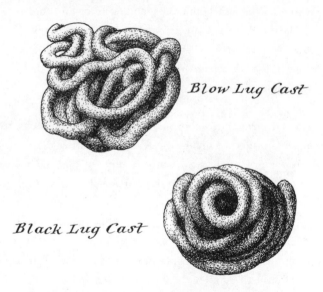

*Blow Lug Cast*

*Black Lug Cast*

Black Lug are typically found much lower down than Blow Lug, often only revealed when tides are nearer springs. All of this is hard to remember, even for zealots, and when things get convoluted, it is usually best to first get ridiculous:

> *If your lug has coiled systematically*
> *And not crapped all chaotically,*
> *Then it can't be a Blow Lug*
> *It must be a Black Lug*
> *And also a very low sea.*

And then to put the sensible hat back on and simplify things:

> *If your lug goes neatly, it's a very low sea.*

WE HAVE LOOKED at the major influences on tidal heights, the moon, sun, wind, air pressure, and temperature, but the water reader can take this to an almost spooky level of investigation, if they choose. The National Oceanic and Atmospheric Administration, a US federal body that covers this area, takes thirty-seven major independent variables into account when predicting tides; that's the sun, moon, and thirty-five others. The oceanographer Dr. Arthur Doodson identified a total of 396 factors at play. It may be a complex area, but tides are an opportunity to wonder at the interconnectedness of the natural world.

# 16

# Water at Night

EVEN THE MOST EXPERIENCED sailors will remember the first time they skippered a boat into a busy harbor at night. It is a time of excitement and beauty, but also of adrenaline and bewilderment. There are just so many flashing and blinking lights out there that the difference between knowing what's going on and feeling a sense of panicked confusion is not as wide as those around the new skipper might hope.

Approaching Cherbourg, the French city on the English Channel, for the first time at night, contending with a strong tidal current, I strained to make sense of the water in front of me. There were green flashes, red ones, yellow ones, white ones, and the broad sweep of lighthouses, too. Whenever I think back to that time, I am reminded of the scene in the farcical film *Airplane II*, where the base commander, played by William Shatner, is brought to the verge of a nervous breakdown by all the lights around him: "I can't stand it any more! They're blinking and beeping and flashing! Why doesn't somebody pull the plug!"

Fortunately, with a little practice, the melee of lights start to make perfect sense, and what once looked like a dark carpet that someone has carelessly thrown a set of Christmas tree lights over, can be deciphered easily, and the dark water around them can be read quickly as a result.

If I had my time again, I would have spent a lot longer appreciating the lights on water at night from the comfort of the shore and from the decks of ferries, before grappling with them as a means of navigating a yacht. In this chapter, I will show you the basics and give you enough of a start that by following some simple principles you will be able to build a picture the next time you look out over a dark sea or estuary.

## THE CODE

Learning to read the code of lights in the water begins with the basics of a new language. The first and most obvious rule is this: When it comes to lights in the water there are no random colors, and every color you see means something. The second rule is that how a light is behaving is very important: Is it on all the time or is it flashing? How is it flashing and how long does it go dark for? Together the color and its patterns of on–off behavior are known as a light's characteristics, and once we take that small step from seeing a sea of random flashing lights to reading the characteristics of each individual light, we are well on the way to reading the dark water.

Starting with the most simple characteristic of all, if a light is on constantly and not flashing at all, this is called a "fixed" light. But if you notice it going dark at any point, things get a bit more challenging and interesting. I have used the expression "flashing and blinking" in the paragraphs above. This expression is commonly and casually used to mean lights that are going on and off, but if you want to crack the code then there is an important difference between "flashing" and "blinking." A light that is flashing can be thought of as darkness broken by short periods of light—it is off longer than it is on—which is a familiar pattern. But a light that is blinking is on longer than it is off, and this is slightly less familiar. A blinking light is formally described as "occulting" and can be thought of as light that is broken by short periods of darkness. Imagine you're in a dark room and there is a flashlight shining at you from the other side; that is a fixed light. If someone starts walking backward and forward in front of that torch in a regular and timed way, that is now a blinking or "occulting" light.

What if the light and darkness times are equal; is that light flashing, blinking, or both? Ah, there is a solution; this light is described as being "isophase."

Some lights flash faster than others, a few change colors, and some even flash Morse code, but we need not concern ourselves with those for now.

## LATERAL MARKS

Lateral marks are the most common lights you will see. They are red or green and are used to mark the edges of channels for shipping. Red marks show the port side (left) of the channel and green marks show the starboard side (right) of the channel. The key thing to remember is that buoys are set out with regards to shipping that is returning from the sea to land, not heading out (this is true for most of the world except the Americas and

Japan, where it is the other way round). When I was taught this I was told that you are more tired, under more stress and therefore need more help when heading home after time at sea than you do when setting out after time on land, so this is the logical way for these buoys to be set out. Whether that is the true logic behind it I don't know for certain, but I mention it here because it has helped me remember it easily ever since.

Lateral marks can have any light rhythm—fixed, flashing, occulting, or isophase. There will likely be lots of fixed lateral marks, but there shouldn't be two flashing ones with the same characteristics within easy sight of each other. The idea here is that the fixed lights are like cat's eyes on a road; they allow a skipper to follow the general course of the channel by sticking between these fixed green and red lights. But there will also be a few important marks that have distinct and unique flashing characteristics, which allow the skipper of a vessel heading into a port at night for the first time to be able to identify not just that they are looking at a starboard or a port mark, but to pinpoint exactly which one.

The way this works in practice is that a diligent skipper will make what is known as a "pilotage plan" before attempting to arrive at a new port. This plan will include a note of the lights they expect to see in succession as they approach their destination, and it will also include a few particular ones that they need to be especially vigilant for. Taking my home harbor, Chichester, as an example, there are long stretches where you will come to no harm by gently pootling along between the green and red fixed lights. But there are one or two places where unwary skippers get into trouble every year because of fast-moving shallow water and shingle or sandbanks in places that might not be expected.

One of these is known as the "Winner Bank." It has three starboard lateral marks alongside it, with their green lights, but since it is such an important seamark for skippers to pinpoint, each of these lights has its own unique flashing pattern. The first flashes once every ten seconds, the second twice every ten seconds, and the third three times every ten seconds. These unique marks usually have their own names too, e.g., "Mid Winner." On any nautical chart these will be clearly marked using a shorthand, e.g., "Fl(3) G 10s" (flashing three times, green, every ten seconds). It is great fun to study these charts and many samples are available online for free, or it may be worth purchasing one for your favorite area, but you don't have to study the charts to get a feel for these marks. Just try to pick out the line of green and red lights marking the passage for the boats and then watch as they follow this line in and out of a port.

After watching for a while you may notice that the smaller vessels, like small yachts, don't follow these channels strictly all the time. There is a good reason for this: These channels are mostly set out for the bigger, commercial vessels, and these big boats will very rarely deviate from these carefully marked channels. But smaller craft can actually avoid the bigger boats, which is especially welcome at night, by sailing just outside the line of the channel, where they can find water that will normally be deep enough for them at most states of the tide. When you see these smaller boats outside the marked channels, it's a sign of a knowledgable local, a careful chart reader or someone who doesn't know what they are doing and maybe about to run aground!

In daytime these lateral marks can be identified just as easily. In the US port marks will be green and starboard marks will be red. They also have their own shapes: Port marks have a boxy, squarish, green shape on top of them, whereas starboard marks have a red, triangle shape on them that points upward.

## CARDINAL MARKS

The next most likely light you will see is called the "cardinal mark," and it is my personal favorite. These lights are very useful for sailors and great fun to look for from land, not least because they follow a logical system that makes most of them very easy to spot and identify.

Cardinal marks earn their name because they refer to the cardinal directions of north, east, south, and west, and use these directions to point the way to safe water. There are four types of cardinal mark, a north cardinal, an east cardinal, a south cardinal, and a west cardinal mark. Each of these is telling the direction of safe water relative to the unsafe water, and it's quite logical: A north cardinal mark means the safe water is to the north of that buoy.

Cardinal mark lights are always white and always flash according to a simple set of rules. These rules work by thinking of a compass and the face of a clock, and however odd that sounds, once you get used to this it will help you to remember these lights easily. I've forgotten the characteristics of a lot of navigational lights over the years and had to relearn them, but I have never forgotten the cardinal mark system. If it seems a bit weird to start with, try to persevere, because once you've cracked it and then gone out to look for them a few times, chances are you will remember the method forever.

If we think of a clock face being divided up into four quarters according to the compass, then we can use the clock to help us identify the lights at night.

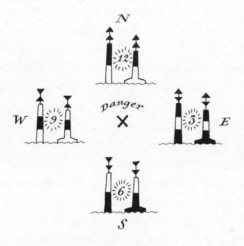

*Cardinal marks.*

Starting with an east cardinal, imagine a clock face and you can see that the east cardinal is going to be over the 3 o'clock area. An east cardinal flashes white three times. Now think of the west cardinal, that is going to be over the 9 o'clock part of the clock. A west cardinal flashes white nine times. The north cardinal is over the 12 o'clock position and flashes white continuously (they probably started with the idea of it having twelve flashes, but realized that busy sailors don't want to be counting that many flashes, so changed it to continuous flashing). The south cardinal would be over the 6 o'clock position and it flashes white six times, but then adds a long white flash afterwards to make it stand out and avoid any confusion with the others.

The cardinal marks are simple and mean one thing—safe water is in the direction of their name—but they are used in a variety of situations. They tell shipping which side to pass, but in doing so draw attention to a wide variety of things, like a notable bend in a channel. They are also often used to "fence off" a potentially hazardous area from shipping, like a reef.

In daytime these marks have their own code, too. They have their shapes on top, formed of two black triangular arrows, and they are painted a mixture of yellow and black stripes as well. But as you will have spotted, little is random in this world of sea marks, and even the stripes have meaning.

A northerly cardinal has two arrows pointing up, a southerly has two arrows pointing down, an easterly has the top one pointing up and the bottom one pointing down. A westerly is the opposite to an easterly, the top pointing down and the bottom pointing up. The black stripe will be toward where the arrows are pointing.

## SPECIAL MARKS

If you see a yellow light, fixed or flashing, then you are looking at a "special mark." These are the vaguest of marks; it's a sort of miscellaneous mark that can be used for anything that doesn't fit neatly into the other categories. They are commonly used in summer for racing, but can be used to mark a wreck on the seabed, pipes, anchorages, marine farms, waterskiing areas, or any number of other things. If you see a yellow light, it is there for a reason, but that reason will take some educated guesswork to discern.

Special marks can be identified easily in daytime, as they are painted yellow all over and usually have a yellow cross on top of them.

## ISOLATED DANGER MARK

If you see a light flashing white twice every five seconds, this is an "isolated danger mark," and as its name suggests, it means that the water is mostly safe in the area, but at that exact spot there is something, like a rock or a wreck, that is a hazard to shipping.

In daytime they are identified with two black balls on top of a structure with black and red stripes.

## LIGHTHOUSES

With hindsight, it is easy to see how history made a perfect cocktail for disaster around Britain in the early nineteenth century. Shipping volumes were growing rapidly in a part of the world that has one of the most intricate, rocky coastlines imaginable, these jagged shores being washed by some of the most powerful currents and extreme tides on Earth.

By the 1830s, more than two ships a day were getting wrecked around the British coastline. Wrecks were so common that they have spawned their own lingo to handle the ensuing legal wrangles, much of which survives to this day. *Flotsam* is the floating parts of a ship or its cargo that has been shipwrecked; *jetsam* is anything that has been deliberately thrown overboard— i.e., jettisoned; *ligan* is wreckage lying on the seabed, often marked by a buoy for retrieval; and *derelict* is cargo lying on the seabed that is unretrievable.

Sadly, the scale and number of disasters had to reach a wincing level before the impetus was found to try and prevent more of them. In our safety-obsessed empirical and secular society, this sluggish response to do the obvious is hard to fathom. Not least since lighthouses have been around for a very long time indeed; the Pharos of Alexandria, which stood over 400 feet tall, dates back to about 260 BC, and the Romans had built a lighthouse in Dover only 300 years after that. But the problem was not so much technical as philosophical. What is the point of building a lighthouse if God is intent on wrecking a ship anyway? This was a view that was held as strongly among the seafarers themselves as anyone on dry land. And the wave-washed corpses strewn on the rocks contributed precious little to the debate.

As Bella Bathurst has recounted in her book *The Lighthouse Stevensons*, the solution was often simple on paper but expensive and technically challenging to realize by stormy seas. Slowly at first, sentiment changed, and by the twentieth century the view held strong that there should be no coastal waters that are frequented by shipping that are out of sight of at least one lighthouse.

We all recognize a lighthouse when we see one by day, and they are equally easy to spot at night, those majestic long beams that sweep the sea and often much of the land, too, before going dark for a period again.

Lighthouses should be thought of in the same way as the lights we have looked at already, but their identification is a little more elaborate and enjoyable. Some have very simple characteristics, but the time period between flashes is often much longer than smaller lights, which is why they can fox those not used to them. With a bit of practice you will be able to nail their characteristics easily, and to do this it is a good idea to enlist the help of some elephants.

As before, it is important to note the color of a lighthouse's light. It will usually be white, but if it is not it will likely be green or red. Some lighthouses have what is called a "sectored light," which is an ingenious and low-tech way of sending a very simple light message to ships. Using filters it is possible for a light to appear different colors or even invisible, depending on where it is viewed from. So it is not unusual for a lighthouse to be invisible when viewed at night from the land as it serves no purpose there, red when viewed from starboard of the ideal channel (remember, in the US this is from the perspective of a ship heading from port to sea), green when a ship is to port of the desired channel and white if it is in the prescribed channel.

Now the fun bit, which comes in two stages. First, count the number of flashes. Do this a couple of times to make sure you've got it right. Next you need to count the number of seconds that make up the cycle. For example, is it one flash every five seconds, or three flashes every twenty? The important thing to remember is that the cycle runs from the start of the flashes all the way back to the start of the flashes again, not from the end of one set of flashes to the start of the next ones (you will get the same time if you go from the last flash to the last flash, but most people find it's easier to go start to start). One helpful trick to counting seconds is to count elephants: one elephant, two elephants, three elephants . . . There are lots of conventions for counting seconds, but most of them rely on three syllables. Elephants have been the one I have used for lighthouses for many years and have served me well.

Repeat the steps above to make sure you've definitely got the light's characteristics, and now if you want to, you can check against a chart, or even just Google the name of the lighthouse or its location if you know them, and see how you did. (Using the Internet to help with learning at this stage isn't *cheating*, it's *accelerating*. It means you'll be up to speed that much faster the next time you're by the water and with no thoughts of the Internet.)

The next step is to try using these skills in anger—well not anger, but on water. If you ever happen to travel somewhere by boat, take a look at a chart or map of the port you are heading to and try to track down any lighthouses near that or near the route you will be following. Then all you need do is find those characteristics and you will be ready to spot them at sea; you will be among the first on board to "see" your destination by recognizing that individual lighthouse.

When you do find a lighthouse on a chart (it will be in capitals) you will see it has small pieces of information about its light next to it. You might see the following on a chart:

Fl(3) 20s 28m 11M

From our look at the smaller lights, you will recognize some of this, particularly the first part. The light is flashing three times every twenty seconds and where no colors are mentioned with regard to lighthouses, we can assume the light to be white. The numbers afterward make sense when you're used to them, too: This lighthouse has a light that is 28 meters above the sea and can be seen from 11 nautical miles away (a nautical mile is 15 percent longer than a normal mile).

If you feel up for a real challenge, have a look at the characteristics of the Needles Lighthouse, at the west end of the Isle of Wight on the English Channel:

Oc(2) RWG 20s 24m 17/14M

Have a think, maybe a coffee, and see if you can crack all of it or some of it, before I reveal the meaning behind this delightful lexicon.

THE NEEDLES LIGHTHOUSE is an occulting sector light, which means that it will be red, white, or green, depending on which sector you are viewing it from. It is occulting, which means it will be on longer that it is off, so in this case instead of flashing twice, it goes dark twice every twenty seconds.

And a nice easy bit, the light stands 24 meters above the water.

It lists two ranges, 17 nautical miles and 14 nautical miles, because it actually has more than one red sector; it has an intensified red light that shines further in its own special sector. Don't worry about that last bit, though; there are sailors who've been at sea for decades who wouldn't be familiar with that, as it's something of a special case.

You might quite reasonably be wondering why on earth we should care how high a lighthouse's light is, since we know from a chart how far it is visible from. Well, this distance on the chart tells us how far the light is visible *from sea level*, but if you are on a large ship or even on land then the distance will be much greater. Using tables, sailors and others can work out how far a light will be visible from if they know both its height and how high they are above sea level. In the Needles example, if you are standing on a ferry deck 10 meters above sea level, you will be able to see the light well before someone in a small boat, who won't see it until they are several miles closer. This is the same concept that we looked at when considering the distance to the horizon in the Coast chapter.

We will leave lights and lighthouses with a skipper's tip. If you are with someone else and trying to confirm that you have identified the characteristics of a light accurately, then there is a little trick worth knowing. A mistake that rookie skippers often make is to ask those near them leading questions and then fail to appreciate that these questions tend to lead to the same, often useless answer.

"Is that light flashing six times?"

"Yes."

"Do you make it fifteen seconds from the first flash back to the first flash?"

"Yes."

"Is that light green or white?"

"It's green . . . wait, maybe white."

People like to please, but that isn't always what is most helpful. Instead ask a question that invites an independent view of things.

"How many flashes do you see?" or "What time do you make it from the first flash back to the first flash?" or "What color is that light?"

## STRANGE LIGHTS

There are many aquatic creatures that give off light at night like the phosphorescent plankton *Noctiluca scintillans*, which glows when disturbed and has earned the nickname "sea sparkle." Most of these creatures are very small, but some are a bit bigger, like the jellyfish *Pelagia noctiluca*. This jellyfish, known as the "mauve stinger," made news in 2007, but not for its ability to glow in the dark. A swarm of them swamped a salmon farm in Northern Ireland, causing almost 1.5 million dollars' worth of damage.

Water readers spend a lot of pleasant time looking out over harbor waters and the boats sailing and chugging through them during the day. The experience is no less enjoyable at night, but if you see what appears to be a phosphorescent flock of small creatures above moored sailing boats at night, then I'm sorry to disappoint you, but you're looking at something less magical. Boat owners hate returning to a long-moored boat to find it covered in bird muck, and one of many tactics to prevent this is to thread a line of CDs over the boat as a bird scarer of mixed effectiveness. The CDs catch any light and bounce it back to our eyes, creating this strange sparkling effect.

## A CELESTIAL DIGRESSION

I was unable to see anything, which made it difficult to roll with the motion of the boat, so I held on tight, gloved hands gripping cold metal. The December air was chilling my face, which made me strangely grateful that the tight blindfold was at least keeping my ears warm. The conversation of the others on the boat was mostly drowned out by the sound of the engines as we headed out to sea. The only clue I could note was that the sea grew slightly choppier, as the distance from land grew.

I was told it was time to take off the blindfold and as I did I could see they were aiming straight at my face. Fortunately this was not a genuine kidnapping; the thing pointing at my head was a TV camera.

A few weeks earlier I had been standing on a beach in Mauritius coincidentally looking out across a dusk sea, when my mobile rang. Cursing myself for not switching it off, I answered it. I didn't catch the person introducing themselves as the line wasn't great, so the first thing I heard was a question.

"If we blindfolded you, stuck you on a boat and took you out to sea, then took your blindfold off, would you be able to tell where you were just by using the stars?"

"Err . . . yes," I replied.

"And if we then gave you an envelope with a secret destination on it, would you then be able to find your way to that destination using just the stars?"

"Yes," I said, then, "who is this?"

"It's the BBC."

The producer I was speaking to turned out to be working for the BBC *Stargazing Live* program, and this call led to me finding myself temporarily disoriented on a boat on a cold, clear night in the English Channel.

Celestial navigation is not strictly about reading water, but it is such a fundamental part of the history of humans on water at night that it warrants a small digression. I can't teach you how to use a sextant in a few pages, but by explaining what I did on that boat you will get a good idea of how celestial navigation works, and this will hopefully add a layer to any time you spend by the water at night. If you have no interest in how to use the stars as a navigator, then feel free to skip this section.

The challenge wasn't to work out roughly where I was, but to pinpoint my position as precisely as possible using only the stars. We could see the sun setting, and I explained to the presenter, Mark, that sunset is a golden opportunity for us to get our bearings; it's a compass in itself (I knew from experience that the sun would set close to 230 degrees that day). I also explained that things were about to get busy. Many people think of celestial navigation as a romantic and leisurely business. At the sharp end, it's not. To be specific, finding direction using the stars can be a laid-back process—it really is very simple. Find the Big Dipper and use that to find the North Star and you are in business.

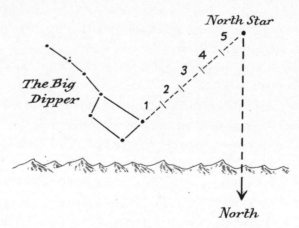

*Finding the North Star using the Big Dipper.*

But fixing your position using the stars is not so leisurely. There is a fairly narrow time window to get the sights you need and when it's gone, it's gone for another thirteen hours in winter. You need enough light to see the horizon and enough darkness to see your stars—twilight is a vague notion to some, but an exact one when it comes to celestial navigation. The reason you need these two things is that celestial navigation is all about angles, and if you can't see either the star you want or the horizon below it, you can't measure how high that star is.

I got us warmed up with some sights of Venus. I called, "Time . . . Now!" And Mark noted down the GMT time to the second. I then gave him the degrees and minutes reading off the sextant. By the time true twilight came we were well drilled as a team and ready.

The director and I had held long conversations by phone in the weeks leading up this moment. Our chats hinged around the method I would use. We settled on a slightly unorthodox approach for a good reason. The usual approach is to take sights of at least three stars, sometimes as many as six. For a program like this it was very important that the viewer was able to follow the logic of what I was doing and to achieve this I was happy to simplify things. I suggested we sacrifice a little accuracy (and redundancy) by only using a two-star sighting and fix. I would get my latitude from the North Star, Polaris, and use a star in the west or east to work out my longitude. It was fortunate that, at the times we were considering, one of the brightest stars of the night sky would be very close to due west.

I pointed out this bright western star, Vega, and everyone on board was surprised that a star was visible when it still appeared to most on board to be late in the day—not early in the night. That is something that surprises many: You can spot stars much earlier than most imagine, if you know where to look. In fact, you can use this approach to find Venus during the day, too. It's worth trying if you haven't before. At sunset or sunrise, at a time when Venus is clear (i.e., reasonably far from the sun), look at where Venus is relative to the sun. Then the following day, if the sky is clear in the middle of the day, look to the same spot relative to the sun (shield your eyes from the sun with one hand) and you may well find Venus again, even near the middle of a bright sunny day.

We took three sights of Vega, then scanned the sky for Polaris. Capella was easy to find in the northeast and then the North Star appeared, very faintly at first. Another three sights and it was time for me to do some arithmetic. In truth, the few astronavigators that remain in the world will normally lean on microchips at this point. I did have that option with an app on my iPhone; it contains data up to AD 2500 and would fill enough books to sink our boat. But both the director and I felt that this would rob the viewer at home of both the logic of what happens next and the romance.

With the help of the *Nautical Almanac* and my sight reduction tables, I began compiling the information with pencil and paper. I explained to Mark that once we had averaged our sights for each star, we would have two key pieces of information for each one. The first was the time of the sighting and the second was a good measure of the angle of each star above our horizon. The first of these, time, was ready for use, because we had used a watch that I had synchronized to GMT earlier that day. But the second, the angle, was not yet fit for purpose. The sums that needed doing before our angles could be useful were simple, but essential.

First, I needed to account for "index error"—this error stems from the fact that no sextant is perfect and nearly all misread by a touch. So long as you know what this error is, you just factor it in at this stage and it does no harm. Next there was "Dip." The Dip stage came as a surprise to all on board. Dip takes account of the fact that you are not taking your sight from sea level, but slightly above it. In our case we stood six feet above a deck that itself was four feet above the sea. It doesn't sound like much, but it means our sights would all be three minutes (one-twentieth of a degree) larger than they would have been from sea level, and this needed subtracting. A small adjustment, but a vital one, as without it we would be out by three miles.

Then there was the "apparent altitude correction," or what I prefer to call the "stick in the pond effect." Whenever you see a star (or the sun, moon, or planets) it is not exactly where it appears to be, because its light gets bent slightly by our atmosphere. The lower a star is, the greater this effect, and sights below 10 degrees are not recommended for this reason. (If you do happen to find a star directly overhead, its light hits our atmosphere perpendicularly and there is no refraction—it is exactly where it appears to be—but that is very rare.)

The final calculations are the real business of celestial navigation. They are not at all complex, but their full intricacies can fill a five-day indoor course. I have never seen the logic of this process explained clearly and I may fail here, but I would like to try.

I would like you to find a lamppost and to stand underneath it. What angle is this light above the street? Answer: 90 degrees. Which means that if you called me and told me that you had taken a "sight" of this light and it was 90 degrees, I could tell you with certainty that you were standing exactly underneath that light. Next, if you took five steps away from the light and gauged its angle above the street, you might come up with something like 70 degrees. The light appears lower, the farther you are from it. And that in a nutshell is almost all you need to know about how celestial navigation works. Let me demonstrate with a strange thought experiment.

Imagine I called you on your mobile and asked you to stand somewhere on the street and tell me the angle that a street light we both know is above the ground. Whatever your answer, I would then be able to gauge roughly how far you were from that light. If you said the light was 50 degrees above the pavement I would say that I think you are twelve paces from the light (it's not magic, just trigonometry).

But—and it is a very big *but*—although I would be able to tell you roughly how far you were from the light, I would not be able to tell you exactly where in the street you were. I would have what is called a "position line," that is, I know the line you are standing on, but it is not an exact point; it stretches in a circle all the way around your light, at an equal distance from it. This is because there is only one set of places in the street where the light appears at that angle above the ground and it forms a circle around the light, with its center at the light.

A single position line is a big clue to where you are, but not definitive enough to be really useful. I would need at least two other pieces of the jigsaw to fix your position precisely. If you told me that when you looked in another direction you could see a second streetlight we both knew and it was at an angle of 30 degrees above the ground, and then turned again to see a third one at an angle of 20 degrees, then I would then be able to pinpoint where you are standing. There can only be one place in the street where light A appears 50 degrees above the ground, light B is 30 degrees above it, and light C is 20 degrees. Each angle creates a line of possible spots where you could be standing and the precise point where these three lines intersect is the only place in the street where you could possibly have observed these precise angles.

This is how celestial navigation works. The stars are our lampposts. The only thing that makes it a little more challenging under the night sky is that the streetlights move—the stars are not stationary relative to our horizon. They rise, set, or rotate because the earth spins. So we need some way of relating their position to time: Hence the tables and chronometers of old and the digital watches and apps of new.

Back out on the English Channel, the angle of Polaris above the horizon told me my latitude, which is after all just a long position line that stretches around the globe. If Polaris is 50 degrees above your horizon, then your latitude is approximately 50 degrees north; you might be in the English Channel, but you might equally be in Ukraine, or Kazakhstan for that matter. Fortunately, my sight of Vega gave me my longitude, which narrowed things down to a precise spot southeast of Weymouth, a seaside town in southern England.

It transpired that I was between 3 and 4 NM off our position by GPS (we probably drifted about a mile during the time it took to take and plot the sights). Which is fair enough and about as good as can be hoped for from a sighting of two stars on a deck that was rolling considerably.

Mark pried open the sealed envelope in his hands. He pulled out a white card, with the word ALDERNEY—one of the Channel Islands—printed clearly on it.

We knew that Alderney was south of us, so all we needed to do was to work out how to head south, which is always nice and straightforward on a clear night. We used Polaris, the North Star, to find north and then used the stars opposite it to hold a course south. As the voyage progressed and the southern stars wheeled clockwise from southeast to southwest, we updated these target stars. This is very similar to the process used by the Pacific navigators and is referred to as following a "star path." The bright star Fomalhaut was ditched early on and then both Markab and Algenib in Pegasus helped us on our way.

Orion rose in the east and Jupiter shone very brightly under the Gemini twins. Hours passed and then lights began appearing over our southern horizon. There was a general glow and many vague splashes of brighter light. There were also three bright and intermittent bursts of distinct light, much clearer than anything else. I pointed up at the southern stars and recapped with Mark. We knew that the southern stars wheel from left to right and we knew how fast they moved.

"So which stars would be a good bet for a southerly course now?" I asked Mark.

"This group, Cetus," Mark replied and pointed at the constellation.

"I agree. Now lower your eyes and look at the horizon directly below them."

"Ah!"

Mark had spotted the four flashes in fifteen seconds that identified Alderney's lighthouse. We'd made it due south to within sight of our destination.

Paper, pencil, sextant, and stars were used to shape our short voyage, but at no point did we turn anything on.

EVEN THOUGH the use of a sextant is outside the realm of this book, anyone who spends time looking at water at night will benefit from befriending the stars. Pliny wrote of the way those who traveled at sea had a habit of noticing how the stars are closely related to where we are on Earth, and it is probably still true that those who enjoy being on or near water take a greater interest in the sky than those who are landlocked. For one thing, if you are on the coast then your night sky will normally stretch down to sea level in one direction at least.

I'd recommend a simple exercise to get you started. Learn to find north using the North Star and the Big Dipper method above, then note how high the North Star is above the horizon. This angle will be almost exactly the same as your latitude. The next time you make a journey south, find the North Star again and note how it is lower in the sky than it is at home. This is the same exercise I was doing on that boat to work out my latitude, the only difference being accuracy, which, if truth be told, is not the fun bit anyway.

# 17

# Shipwatching

*So we were able to see the art of these captains and sailors in managing the jalbah among them; it is extraordinary how they bring it through the narrow channels and lead it among them like a rider on a horse which is sensitive to the rein and easy under the bridle; and in this they show marvellous skill, difficult to describe.*

IBN-JUBAYR, C. TWELFTH CENTURY

THE ALLURE OF BOATS working their way close to shore has the power to hold our gaze. The experience can be further enriched when we know what a trained eye will look for. Once upon a time the shape of every boat could reveal truths about the local waters—the Stroma Yoles of northern Scotland were solid, beamy boats with a curved stem that allowed them to ride the local waters well and then run up the beach afterward. Sadly, such locally perfected designs grow rarer each year. True to the rest of the book, this chapter will focus on how certain clues and signs help us to interpret the water, not the boats themselves. There are techniques that can be employed from the slightest sloop to a supertanker.

If the word "sloop" in the line above made you feel the first pangs of dread that this is an exclusive world, then you are far from alone. The language of sailors has evolved in its own briney world for so long that it can sound like a closed shop. Some are won over by the romance of this lingua franca, but there is no need for anyone to be deterred by it. The best thing to do when the words turn nautical is laugh and think of the sailor who tells the dentist which tooth is causing him problems: "'Tis the aftermost grinder aloft, on the starboard quarter." The trick is always to remember that language cannot change the physical reality of what we are seeing and sensing. Think of a sailing yacht in the distance; she may be "beating into a stiff breeze, close-hauled on a starboard tack and well reefed," but that cannot change what the wind and water are doing—they are only labels. All we need to do is remember some very simple rules and all the ships under sail we see will yield clues about the water around them.

Firstly and most basically, the windier the conditions, the less sail a boat will put up (*carry*). So in light winds expect to see a lot of sail up and in stronger winds notice how the sail area is reduced (*reefing*). Next, notice how the closer a boat is pointing to the direction the wind is coming from, the more closely the sails will be aligned with the boat itself. If you draw an imaginary line from the front tip to the back of the boat (*bow to stern*), the closer the sails appear to be aligned with this line, the closer to the direction of the wind the boat is trying to sail.

Let's look at two extremes. If the wind is blowing from directly behind a yacht, then the sails will be as far out as safely possible; they will be almost at a right angle to the line of the boat. This is called *running*. No boat can sail directly into the wind—it defies the laws of physics—but modern boats can sail fairly close, perhaps 45 degrees off the direction of the wind. However, in order to achieve this feat the sails have to be pulled in tightly toward the line of the boat. This is called being *close-hauled*. There are points of sail between these extremes, like a *beam reach*—when the wind is blowing perpendicular to the direction the yacht is heading:

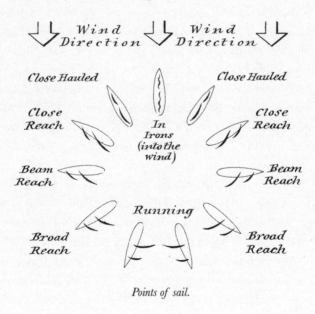

*Points of sail.*

If you are new to these ideas, instead of being put off by the language, tame the whole thing by viewing every yacht you see as your personal weather vane. Once you have practiced spotting the various sail patterns out there (and they apply from a tiny dinghy to a superyacht), you are ready to spot how these two basic ideas of *reefing* and *point of sail* work together. Imagine that there is a breeze blowing at 15 knots (Force 4), and there is a yacht that sails on average at 6 knots. One of the biggest surprises for those new to sailing principles is what a massive difference the "point of sail" makes to how windy it feels for that particular yacht. This is a difference known as *apparent wind* compared to *true wind*, because even though the wind strength over the water itself may not change, the wind strength appears to change enormously for each boat depending on whether that boat is trying to sail into the wind or with the wind behind it.

Let's look at this idea in practice. In this example imagine our yacht is sailing with the wind; it is *running*. You spot this easily because its big sails are well out, near right angles to the line of the boat. The wind those on this boat feel will be the true wind minus the speed of the yacht, 15 knots minus 6 knots, which equals 9 knots—a gentle breeze. No wonder they need as much sail up as possible to go anywhere at all. Now, imagine this same yacht turns until it is sailing almost into the wind and is now close-hauled—the sails have been hauled in tight to the boat. The wind that those on the boat feel and that now fills the sails is radically different; it is now closer to 15 *plus* 6—21 knots. This is a stiffer breeze; the apparent wind has more than doubled, and probably doesn't require as much sail up for the boat to speed along, so the skipper may decide to reduce the sail area and reef.

Hopefully you can now see how the angle the boat is moving relative to the wind, the angle of the sails relative to the wind, and the amount of sail up are all interrelated. You won't have to watch boats for long to spot all of these things happening. In busy places, you're likely to be able to spot boats on all points of sail at the same time.

And then, with a final easy tip for something to look for, it won't be long before you are suggesting that a skipper reef their sails. If a boat has too much sail up for the wind it is experiencing, one of the symptoms is that the boat will lean over (*heel*) too much. This makes the boat inefficient, as the hull shape is not designed to sail most effectively when it is leaning far over. So in most conditions the skipper should reduce sail to stop this heeling.

There are exceptions, and if you are looking at racing yachts then all sorts of odd things will be going on that may or may not lend them an advantage. But as a general rule if you see a boat sailing with the wind behind it, running nice and calmly, and then notice it turn toward the wind and lean right over, heeling dramatically, this is a sign that the skipper has not fully anticipated the difference between the true and the apparent wind out there.

Bringing it all together, there are three clues to look for in sails: the amount of sail up, the angle of those sails relative to the boat, and whether the boat is leaning over. If you practice looking for these three simple clues for a while, you will notice that your appreciation of what the wind is doing over different patches shoots up very quickly. In short, the sailing boats act as wind vanes, which give us a much fuller picture of the local wind, and this helps us read the water.

There are other clues to the water to be found by studying sailing yachts, even ones that are not moving. If you see a shape resembling a black ball hanging somewhere near the front of the boat, you are looking at a yacht at anchor and therefore relatively shallow water. This is one of many "day shapes," visual signals that boats are legally obliged to display to other vessels to communicate what they are doing (it may be a legal obligation, but it is a waning habit, and as a general rule commercial vessels in the developed world still follow it and British leisure sailors the world over often do, but between those two it is a bit of a lottery). The "at anchor" day shape is the only one you are likely to see regularly, although there are a host of others and I'll mention three more, as there is a beauty in their arcane nature.

A dark cone pointing downward means a sailing vessel that is using its engine (only important because the rights of way for a sailing vessel change when its engine is used). Three dark balls, one on top of the other, mean that the vessel has run aground. And my personal favorite, just for its surreal blend of traditional elegance and contemporary horror: three dark balls, one at the masthead, and one at each end of the foreyard means . . . a vessel engaged in mine clearance.

The International Code of Signals is the accepted means of sending messages at sea without resorting to spoken language. A combination of flag designs, associated letters, and Morse code mean that one vessel can send a message to another by radio, by flying a flag, by flashing light, or even by loud hailer, even if they have no idea of the nationality they are communicating with. For example, a flag with a small red square surrounded by a white area all inside a blue border means "I require medical assistance." The same thing can be said by flashing the letter W in Morse with a light: dot, dash, dash. Most of these signals are rare, but one that is seen surprisingly often, when you know to look for it, is the "Alpha" flag.

The Alpha flag is a white and blue flag; one half is white (nearest the pole) and the other half is blue and looks like it is missing a triangular bite from it. It is very common to see it flying from a rigid inflatable boat, one that is bobbing not far from the shore, and identifying that as the dive boat. This flag means, "I have a diver down; keep well clear at slow speed." Look nearby and you may spot a buoy in the water with a red flag with a diagonal white line across it, also known as the "diver down" flag and marking the water that the divers are operating in. It's not surprising that these two flags are some of the few signals to remain in constant use, as divers dread speedboats zipping through the area they plan to surface in. If you pass any naval boats or facilities, have a look for any combination of these flags as they are still in use on many of them. There are flag combinations that mean that gunnery is taking place and another that means that submarines are maneuvering in the area.

There is one flag that always amuses me, one that I think must have been designed more for communication in relationships than anything nautical: the X flag, a blue cross on a white background; it means, "Stop carrying out your intentions and watch for my signals."

At the beach the lifeguards have their own and, you'll be relieved to read, much simpler code of flags. Red means don't swim and a pair of half-yellow, half-red flags mark the area that is recommended for swimming, as it is monitored by the lifeguards. A pair of checkered black-and-white flags mark the "surfboard and watercraft zone" and an orange windsock means that there is a significant offshore wind, so care should be taken with inflatables.

IF SAILING YACHTS remain the aesthetically appealing side of seafaring, then cargo vessels have grown to become the inescapably brash side. The great container ships may lack delicacy, but these behemoths do still offer some good water clue hunting.

The first thing to do is watch how the ships come and go along pretty much any coastline. You will spot how regular their routes are, all these cargo vessels follow designated shipping channels when close to land. Out on the ocean they can be flexible, but in busier waters they are corralled into lanes, just as we are in our cars. And like roads, these channels are also segregated by direction of travel, so the next thing to work out is which channel is closer to you: Is it the shipping heading to your left or right?

The relative size of the ships may solve this for you, but if that doesn't work then there is a trick for solving the mystery: the paler ships are further from you. An optical effect known as "Rayleigh scattering" means that the farther away something is then the paler and nearer to white it will appear; the colors effectively leach out as they pass through the atmosphere. Passing ships often appear a similar distance away—they may even appear on collision course—but with practice you will notice how one is slightly paler than the other, and this is the more distant one. If you have time you can then check to see how you did by waiting and watching to see which ship passes in front.

Closer to land than these main shipping channels, there is usually frenetic motor traffic of a smaller variety, traveling locally and using the inshore waters, but look beyond them and you'll spot the cargo lanes.

If you spot any cargo vessels close up, it is well worth looking at their hull. It did not take sailors long to work out that a ship that carries too much may be vulnerable in heavy seas, but sailors were rarely the ones to make the decision about how much cargo a ship could safely carry. The merchants making the profit would have had a different view from the deckhand, especially if the trader never set foot on the vessel. This led to a wrestle between greedy traders and cautious captains that lasted centuries. The first attempts to regulate how much a ship should carry go back over four thousand years to ancient Crete, but it was not until the worrying loss of ships in the nineteenth century that an attempt was made at a more systematic approach.

The problem was ingeniously reinvented as the solution. To appreciate the beauty of this solution it is well worth doing a simple experiment at home. Fill a sink or basin with water and then float a cylindrical glass upright in it. Most upright empty glasses will float, albeit in an unsteady and precarious way. If you add a little water to the glass and then set it afloat again you will notice that it is now much steadier. A little ballast low in a boat is a lot safer than none. If you recreate the sea and make some very gentle waves in your basin the glass may bob a bit, but shouldn't sink.

Now visually gauge the distance between the surface of the water in the basin and the top of the glass. In this experiment the glass represents the hull of a cargo vessel and the distance from water to the upper edge of the hull is known as the "freeboard," i.e., the dry bit of the hull. Depending on your glass and the amount of water you filled your glass with this freeboard height will vary, but one thing is certain: If you add more water to your glass this height will decrease. As the freeboard decreases the vulnerability of the ship to rough seas increases, and this is exactly what has been happening in cargo ships for millennia.

If you now fill your glass until the freeboard is very small and then try creating some waves in your basin, one of these waves will eventually lap over the edge of the glass, our hull, at which point a vicious cycle is quickly set off. More water is now in the glass, which means a lower freeboard, which makes it even easier for waves to get in and very soon the glass, the ancient trading vessel, and the one built last year all sink very quickly and with a klip-plunk.

Samuel Plimsoll, a nineteenth-century English politician, realized that a low freeboard height can present a problem, but he also appreciated that it becomes the solution if we take a very keen interest in it. In other words, we can tell if there is too much cargo in the boat by looking much more carefully at how high the water rises up the side of the hull. And the easiest way to do this is by drawing a ruler on the side of the ship, calibrated according to an architect's or engineer's understanding of that boat. These lines, which became known as Plimsoll Lines, were such a simple and brilliant success that they became law and proliferated around the world. To this day we can look out for them on a wide variety of ships.

Like so much of the nautical world, there is a simple code to be broken before we can enjoy reading Plimsoll Lines. There are usually two main parts to these lines. There is the vertical ruler, which is the key part, and alongside these vertical markings you will usually see some letters, like, TF, F, S, W, WNA. These letters are abbreviations for water types, Tropical Fresh, Fresh, Summer, Winter, Winter North Atlantic. Everything becomes more buoyant in salt water than it does in fresh water, and the temperature of the water affects its density, too. This means that a boat that is perfectly loaded for cold salt water will sit noticeably lower, and potentially more dangerously, in warmer fresh water.

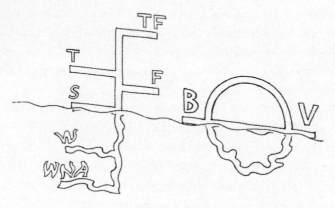

*A Plimsoll line.*

The authorities take a more detailed and forensic view of these marks, but we can satisfy ourselves that any ship with water close to the top of these lines is very heavily laden, and as the waterline drops and the lines rise the lighter the load she carries. Some have observed wryly that we can gauge the global economy by studying these lines and that shortly after 2008 they leapt out of the water. (Incidentally, at times of global recession these cargo ships quickly exceed the capacity needed for trade, which is why we will see so many of them idly lining deep anchorage areas around coasts, for months or even years at a time.)

Alongside the main Plimsoll Lines, you may spot a circle with a horizontal line through it, accompanied by a pair of letters. These don't matter in terms of understanding the water, but to complete the codebreaking you might like to know that these refer to the authority that certified the lines and are therefore a clue to the origin of the ship. LR—UK (Lloyd's Register), BV—France (Bureau Veritas), AB—American Bureau of Shipping, NK—Japanese (Nippon Kaiji), etc.

## WATER TRACKING

After bad weather had forced us to run for shelter on the dhow in Oman, I walked to the edge of a small, rocky perch and watched the storm approach. The waves changed direction beneath me as the new, stronger winds imprinted themselves over the old ones and I watched closely as they bent toward the land. Then I found myself giggling with the locals as they danced. Heavy rain is rare enough in this part of the world to wash in a festive atmosphere.

The following morning I scrambled up into the jebel, the hot, dry hills that looked out over the sea. I took with me a rucksack with basic food, some water, and a vague plan. I wanted to study one particular relationship between the boats and the water, one that has intrigued me for years.

We are all familiar with the idea that boats leave wakes in the water, and the bigger and more powerful the boat, the bigger the wake we expect. If you've ever bobbed about on a river in a small boat, you can't help but develop a sixth sense for the type of boats that stir the water up most and their oblivious owners, the ones who take a sip of their drink as their wake rocks you one way, and another sip as the reflection off the banks sends you the other.

Many years ago, sitting on a hillside in Greece, resting in a shady spot during a hot walk, I witnessed something that pleasantly shook my understanding about wakes and water. A small motorboat swung in to the large bay I was looking down into; it had a look around and then it swept back out again. The white foam of the wake washed out toward the edges of the bay, bubbles rose in a line behind the boat, and then it was gone. I sat there, drinking water for a while. Then I pulled my wide-brimmed hat down over my face, leaned back against a rock and considered going to sleep.

It was the middle of the day and too hot for that, so after a few minutes I stood up again, waited for my eyes to adjust to the bright light of the sun on the rocks and water of the bay, and took another sip of water. Something in the water struck me as bizarre. There was a perfect curve in the water, one that I recognized immediately; it was the course that the boat had followed, about ten minutes earlier. There were tracks in the water.

One of the oldest human skills is tracking—understanding the routes, timings, and behavior of the animals and people who have passed through an area before you. It is a skill that enabled our ancestors to hunt better and avoid being hunted by competitors, and as such it was a vital skill for thousands of years. It has seen a renaissance in recent times, as more and more people have appreciated the joy and satisfaction to be found in applying ancient and fundamental wisdom. But I had never come across the idea of water tracking, so from that moment on a Greek hill, I have been intrigued by its possibilities.

One of the universal truths of human observation is that we see more of what we expect to see and less of what we don't expect to see. This strange and simple fact has huge ramifications for anyone interested in outdoor clues of any kind. The sheer wealth of detail from our senses—especially our eyes—that our brains are trying to process means that anything that our brain can automatically decide is irrelevant it will happily do so, without troubling our often clogged-up conscious mind. It's strange how the the brain doesn't register a second "the" when written next to the first "the" in this sentence.

Much of my work is not about teaching people to see things that are hard to see, but in showing them how to notice the things that hide in plain sight. Once you come to know that the orange lichen, *Xanthoria*, is common on roofs and tree bark, but much more common on the sunlit south-facing sides, it surprises many people how this bright-colored organism, one they often pass every day of their life, has "hidden" itself so well. And nothing delights me more than when I stumble on a way to turn this logic on myself. The water tracks in that Greek bay were one such moment. From then on, I began to notice that every object that moved through the water left a much longer signature of its path through the water than I had ever imagined or noticed before.

Once we become sensitive to a new detail like this, it often appears to crop up everywhere, in life and in literature. So I was surprised and delighted to discover that not only had the author of *Waterlog*, Roger Deakin, noted his own wake patterns while swimming, but that Tarka the Otter was leaving them, too, even when swimming underwater. It turns out that submarines at shallow depths also leave a noticeable wake at the surface.

My inquiries led me into the science of wakes, and that quickly becomes a foreboding land of equations and competing theories. Lord Kelvin, the formidable British physicist, did make one mathematical discovery that is helpful to know and use whenever we see wakes. A boat's wake will spread until it is close to 20 degrees from the main track of that boat. Since it spreads in both directions, this means that it is 40 degrees from one wake wave to the other, and since an extended fist makes an angle of close to 10 degrees for most people, this means that if you are on a boat looking back, then it will be four fists from the wake wave on the left to the one on the right, whatever boat you are on, however fast it is traveling and however far you look into the distance. Amazingly the same angle will apply to ducks paddling through water and even to a stick drawn through water.

In many ship wakes, there will also be a set of waves created between the main wake waves and roughly perpendicular to them. These are known as transverse waves, and each one forms part of a circle, the circle growing bigger (and therefore the curves growing shallower) the farther from the boat.

As is always the case, the appearance of these waves changes depending on the direction of the wind, and on a windy day by a busy river like the Thames, you will be able to spot how the wakes from similar boats passing in opposite directions look different.

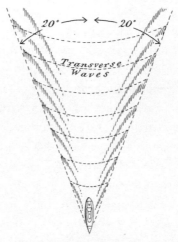

*A boat wake and Kelvin's dependable angles.*

My fascination since that sighting in Greece has lain not so much in the patterns of disturbance in the immediate aftermath of a ship passing through water but in the way that the ship leaves a more lasting mark on the water in its path. Directly behind the ship there is an area where the water that has been whipped up by the propellers rises to the surface. It is easy to spot behind most boats as a different color, usually with lots of bubbles rising to the surface, too. Soon the commotion in this line of water calms down, but the water does not return exactly to its previous state—it retains a peculiar glassiness. This has been observed for many years and has a few nicknames, like the "dead wake." This dead wake remains noticeably different from the water around it for far longer than I would ever have suspected, and depending on the size of boat, its speed, and the natural water state, it is possible to spot the tracks of boats long departed.

Peering out over the Omani jebel that day I saw tracks of small fishing boats that lasted at least seven minutes after they had passed out of sight, and one vessel from the Omani navy left tracks that were easy to read twenty minutes after it had departed. Wake waves from larger vessels travel farther but don't leave longer-lasting tracks; it is the dead water that their great propellers churn and then iron that is visible for longer.

Looking for these delicate tracks in the water forces us to sharpen our gaze and heighten our awareness, and whenever we do this we are likely to spot other signs, too. Staring down over the brown, hot rocks in Oman I noticed a set of what appeared to be boat tracks that lasted too long to be true. Time proved them to be the gentlest of currents sweeping around the headland I was perched above.

# 18

# Rare and Extraordinary

IN THIS CHAPTER we will look at some water phenomena that you are unlikely to see often, but are still worth
knowing about, as they will have a big impact on you if you
do come across them. The dramatic effects of some of these
phenomena also mean that they have garnered more attention
and get talked about a lot more than some of the subtler effects
we have looked at so far.

## KELVIN WAVES

The earth's rotation sets up a phenomenon in the oceans known
as "Kelvin waves." The detailed science is mind-boggling, but
the basic principle isn't too bad. When you stir something like
a cup of tea it is easy to see how this motion creates higher
areas and lower areas, little waves that swirl around the cup.
The oceans are being stirred gently, pushed by the earth as it
rotates, and this sets up these Kelvin waves.

Since Kelvin waves travel one way, from west to east, it means
that the aspect of the coastline is an important consideration

when it comes to tide heights, because the west-facing coastlines tend to bear the brunt. Kelvin waves explain why a west-facing port will often see more dramatic tides than an east-facing one, even if they share the same patch of sea.

## TSUNAMIS

For many decades the word "tsunami" and "tidal wave" were used almost interchangeably by many people, but this was unnecessarily confusing. Things were tragically clarified in 2004 when the Indian Ocean tsunami killed hundreds of thousands and forced the world to confront and better understand these aquatic behemoths.

A tsunami has nothing to do with tides; it is an oceanic wave created by a massive underwater disturbance, normally an earthquake or volcano. This seismic shock creates a set of waves that are surprisingly long and low when they begin life—only about one or two feet high, with up to ten minutes between crests. In 2004, for the first time accurate measurements were made of a tsunami's height in this early stage, using radar: two hours after the earthquake, the waves were gauged to be two feet high.

These still gentle waves then radiate from the area of their violent creation, moving at great speed, 500 miles per hour being typical. And like all other waves, their height actually decreases slightly as they spread out over the open ocean. Three hours and fifteen minutes after the 2004 earthquake, the waves were down to just over a foot.

There is no better or more terrifying demonstration of the physics that dictates that a water wave's height and length will change when it reaches shallow water than a tsunami. When these long, low waves reach the shallowing waters of a coast, the waves shorten and rear up to devastating heights. 230,000 people in fourteen countries lost their lives in December 2004, when waves that were once lower than knee-height were transformed into 100-foot killers that swept through coastal communities.

One of the few signs from shore that a tsunami may be approaching is the sudden receding of the water as it is drawn out to sea into the growing wave. The Moken sea gypsies of the Andaman Sea noticed this and were some of the few to appreciate the gravity of this sign and get out of harm's way in time. Saleh Kalathalay, a spear fisherman, noticed an unusual silence in the animals like the cicadas and ran around to warn those in the village. The villagers were persuaded by the signs and moved to higher ground. The tsunami destroyed their village, but they were saved.

## WHEN THERE IS NO OTHER EXPLANATION FOR A WAVE . . .

If you watch an area of shallow, coastal water over a long period, you are likely to see several of the different wave types that we looked at earlier. There may be some created by swell arriving from a long way off and at times there will be plenty of waves generated by local winds. You will also sometimes see waves arrive that have been generated by the wakes of boats, near and far.

But what if there is no swell, no wind, and no wakes? Is it still possible that waves will arrive? Yes. As we have seen, the tides can be thought of as very long, low waves that encircle half the planet. But we have also seen how waves will shorten—their wavelength decreases—whenever they arrive in shallower water.

It is occasionally possible to spot waves arriving on a shore that have their origin in the tides rolling around the planet. These are true tidal waves and although quite common, they are very rarely recognized for what they are.

## TIDAL BORE

If the tidal waves mentioned above arrive in a place that funnels this energy and it is then met by the friction of a channel, a serious wave begins to rear up. If this turbulent new wave then has to fight an outgoing current, the result is a lot of pressure building up, which is finally released as a surging wave of great force known as a "tidal bore." The topography and tidal rhythms are dependable, and so tidal bores tend to happen with predictability in certain places and not in others. Cook Inlet in Alaska has a tidal bore, one that leaves fish along the banks after the wave's gone.

The River Seine in France also has one, known as *le mascaret*, and in the days before its behavior was well documented, understood, or predicted, it swept Victor Hugo's newly married daughter and her husband to their deaths.

## AMPHIDROMES

All the forces that make our tides go up and down cancel each other out in some places, leaving a sea without tidal highs or lows. These places are known as "amphidromes." The currents still run in these places, but there is no vertical movement of water. Amphidromes are normally well out in the sea and have little impact beyond being a curiosity.

## ROGUE WAVES

In February, 1883, a substantial steamship, the *Glamorgan*, was hit by a wave that dwarfed the ship and smashed over it, tearing away a mast and destroying both the deckhouses and the bridge. The 320-foot ship sank the next day, but this allowed enough time for 44 crew to escape in lifeboats and tell the tale of a monstrous wave, one that no ship had ever been built to cope with.

For decades, scientists thought reports from seafarers of waves that were too big to be believed, waves that dwarfed their neighbors and swallowed ships whole, were fanciful sea talk and little more. The thinking was that the mathematical models that described almost everything that was going on with sea waves made no allowance for rogues and giants, therefore these giants couldn't exist. Hundreds of eyewitness reports did little to challenge the wave equations that the mathematicians had developed.

It turned out that the scientists were relying on a slightly simplistic view of the way sea waves behave. It was a view that changed on January 1, 1995, when a wave hit a gas industry platform in the North Sea. It was a wave that simply couldn't be made to fit the scientists' models. This massive wave was measured by laser sensors to be 84 feet (25.6 meters) high, more than double the height of the other, already huge waves in the area. Some small damage was done to the gas platform, but the math that had governed our understanding of what waves were possible was smashed to pieces.

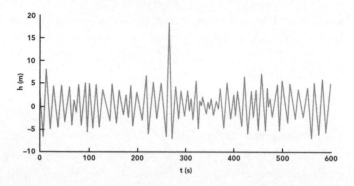

*The giant wave, measured from an oil platform in the North Sea on New Year's Day 1995, that confirmed the existence of "rogue waves."*

On average the waves that the wind creates on the sea fit very neatly into certain size patterns, and these normal waves also fit neatly into mathematical models. As we have seen, wind strength, distance, and length of time will create waves of a certain height *on average*. But the key historical misunderstanding was that these are only guideline probabilities, and waves that don't fit this predicted size do happen—they are just not that frequent. It is all a question of probabilities; a hundred waves of similar size may go by, but that doesn't change the fact that there remains a small probability that a much larger one will go by next. And as the ships' captains, the ones that survived at least, had known for centuries, there is a tiny probability that a monstrous wave will be created every so often.

Some of the factors we have looked at earlier may increase the likelihood of both big waves and also a rogue wave; strong currents running against big, wind-driven waves, for example. This is probably why certain places record more sightings and mysterious shipwrecks than others, the Cape Agulhas area off South Africa being the best-known. Other than avoiding these areas and storms, which may not be an option for some seafarers today and certainly wasn't for many in the past, there is little that can be done at the moment to predict exactly where and when these waves may strike shipping. It is a little ironic that the latest science lends a little credibility to the traditional fatalistic sailor's philosophy that when your number's up, it's up, and that's just that.

## THE PLUGHOLE PUZZLEMENT

Large vortices that spiral inward will rotate counterclockwise in the northern hemisphere and clockwise in the southern hemisphere. This direction of rotation is due to the Earth's rotation, causing something called the Coriolis Effect. On a large scale this motion is very dependable and allows forecasters to predict the behavior of low pressure weather systems and oceanic currents. This is guaranteed to happen in large systems like the weather or ocean, but scientists have long puzzled about how small a body of water could be and still demonstrate this effect.

It turns out that the scientists largely agree that water will behave in this way all the way down to . . . the bathtub, but here they disagree. Some, like the French hydraulic engineer Francis Biesel, claimed that a bath was too small to demonstrate this effect and that the water would spiral around in either direction anywhere in the world. But the American engineer Ascher Shapiro was not ready to have his bathtime fun taken away and so set up a controlled experiment to answer the question once and for all. Shapiro claimed, in 1962, that under laboratory conditions at the Massachusetts Institute of Technology in the US (and therefore in the northern hemisphere), water in a bath will rotate dependably counterclockwise as it falls down the plughole. His results were not universally accepted. What everyone agreed was that the initial conditions are vital to the outcome: The water needed to settle for at least twenty-four hours to avoid any residual motion.

To demonstrate the sensitivity of the direction of movement in the water as you pull the plug, all you need to do is give the water near the plug a quick stir with your hand before pulling the plug. Whichever direction you choose, that is the direction the water will spiral as the bathwater flows out, and it doesn't change direction, regardless of which hemisphere you are in.

A fair summary of the bath situation is this: Water will rotate counterclockwise because of the Coriolis Effect in any large system in the northern hemisphere, but the smaller the vortex, the more likely it is that initial conditions will influence the direction of movement you see. You can predict the direction of ocean vortices with near certainty, but for you to be confident of predicting the Coriolis Effect in tiny ones like your bath, you would need to turn your bathroom into a sterile laboratory where there was no air moving and then wait at least a day before taking the plug out, ideally using a robotic arm.

## WATERSPOUTS

I will never forget flying back over the English Channel in a light aircraft many years ago. On an otherwise sunny day, I noticed that there was a towering and ominous cloud a few miles south of the Isle of Wight. Something about the cloud made me uneasy, and I studied it for a long time before my eyes made out the thin, whirling column of water below it. I had seen my first and to date only waterspout. I banked away from the cloud, radioed air traffic control to report it, and they promised to pass on the sighting to other air traffic and the maritime community.

There are two types of waterspout, tornadic or non-tornadic, the latter being what I had seen that summer's day. Non-tornadic waterspouts are very localized and commonly seen on otherwise good-weather days. They have short lives, and although wind speeds can be serious in spots, they do not cause widespread damage. Tornadic waterspouts are, as the name would suggest, tornados over water and are totally different beasts, presenting a very serious threat over a wide area.

## OCEAN GYRES

The Coriolis Effect bends the major ocean currents clockwise in the northern hemisphere and counterclockwise in the southern hemisphere. When this effect is combined with the way the oceans are hemmed in by the continents, the result in some places is an ocean gyre—an enormous, swirling mass of water.

The gyres act as flotsam traps, as both natural and man-made floating debris get trapped in these rotating systems, often for years. This has led to an unfortunate concentration of rubbish in some places; one zone has been nicknamed with shame as "The Great Pacific Garbage Patch."

## WINDROWS AT SEA

Traditionally the word "windrow" is used to refer to the long lines of cut crops we see in fields at harvest time, but the word has been borrowed and is applied to lots of things that form long lines in landscapes.

Wind blowing over open ocean can set up a phenomenon known as "Langmuir circulation," where vortices of water form, like corkscrews in the ocean. These corkscrews form long lines parallel to the direction that the wind is blowing. The vortices force the water to rise in long lines in places and sink in others. The net effect is that long, visible lines form in the ocean that can consist of calmer patches, rougher patches, seaweed, or other flotsam.

If you see long, straight lines out at sea and they are in line with the wind you are probably looking at windrows. (But remember that if you see lines of glassy, calm water very near the coast on a day of light breezes, you are probably looking at slick lines among ripples, see the Reading Waves chapter.)

## WHIRLPOOLS

If strong tidal currents encounter the right topography, then these fast waters will be set into a spin called a "whirlpool." The greatest whirlpools can be heard a dozen miles away. The Corryvreckan off the island of Jura in Scotland is one of the world's biggest and most powerful whirlpools and it nearly drowned George Orwell in 1947. And off the coast of Maine and New Brunswick, Canada, Old Sow—the western hemisphere's largest whirlpool—can be as large as 250 feet in diameter.

## SLIPPERY WATER

Sailors at the 1968 Acapulco Olympics experienced a tricky phenomenon known as a "slippery sea." From grand estuaries to more modest rivers, it is likely that you will spend some of your coastal time near the point where river water meets the sea, not least because these form popular spots.

When the tide is flowing out, there will be times when fresh water flows out into the sea. Freshwater is more buoyant than salt water and so the two don't always mix well the freshwater may form a layer on top of the brine, especially if it is at all warmer, which is likely in summer. This slippery layer of water then behaves in a noticeably different way from the seawater beneath and all around it. This slippery layer will get driven by the wind and can lead to it moving in a different direction from the salt water. The sailors at the 1968 Olympics needed help from strategists to work out why neighboring patches of water were moving in such counterintuitive directions.

## WILL-O'-THE-WISPS AND JACK-O'-LANTERNS

In some still water, like swamps, the amount of dead organic matter that bacteria are feeding on beneath the surface leads to the water running out of dissolved oxygen. The bacteria don't give up at this point, but they are replaced by anaerobic specialists that break down the decaying plant and animal matter using different processes. Anaerobic bacteria produce methane gas, and if enough of it is produced then it bubbles to the surface. Very occasionally this gas will spontaneously ignite, creating dancing blue flames at the surface of water. These flames have been given a number of nicknames, including "Will-o'-the-Wisps" and "Jack-o'-Lanterns." They remain on my "would absolutely love to see one day" list, and I hope you have more luck spotting them.

## THE GREEN FLASH

Spend enough time around nautical folk at cocktail o'clock and before long you will hear someone discussing the "green flash." When atmospheric conditions are right (and especially during a meteorological phenomenon known as a "temperature inversion"), there is a moment just after sunset or just before sunrise, when the red and yellow parts of the sun's light cannot bend over the horizon, but the blues bend too much and get totally scattered in the atmosphere. At this moment, there is only one color that is able to make the journey from just below the horizon to your eyes, and if you are patient and lucky, you will see a burst of bright green at the horizon. The green flash is rarely spotted and has a semi-mythical status, but it is a real phenomenon and can be seen.

If you happen to notice that the sun is stretched a little vertically, as opposed to the much more common squashing effect, then this is the sign of a temperature inversion, when warm air sits on top of cold air, and offers ideal viewing conditions for spotting the green flash.

## FLYING FISH

There is something about being alone on a boat out on an ocean at night that perfectly blends delight and fear. There is a sense of wonder at the vastness of the seas and starry heavens, tinged with a realization that you are at the mercy of nature, so it's best to enjoy good, calm times while they last. I was in this situation on the Atlantic in December 2007, admiring Mars and Orion, listening to the reassuringly constant slapping of the bow through modest waves, when I received a shock I was unprepared for.

One of my dawn routines was to walk the decks of the small yacht, inspecting the sails and rigging and casually picking up any flying fish that had landed there the night before and returning them to the sea, mostly long dead. I did not cook and eat them, as Thor Heyerdahl's crew did on their raft, but did take some comfort from the fact that it would be hard to starve in a part of the world where edible fish leap out of the water and pretty much into the pan. One night a flying fish made its way back into the sea alive with my panicky help, but not before it had hit me hard and fast in the face, triggering a scene of comical flailing by both the fish and me and a full thesaurus of swearing.

Ever since this abrupt personal introduction, I have held a small fascination with flying fish. There are believed to be between sixty and seventy species, divided into two- and four-winged groups, and they can grow up to a foot and a half in length. Technically they don't fly, but glide with fixed wings, covering between 50 and 300 feet with each glide. The record was set in Japan, where one was filmed gliding for forty-five seconds. That may not sound impressive, but compare it to humans who cannot take off out of the water and glide for half a second.

Their flying is an escape tactic and is ingenious, not just because of the speed it gives the fish or the way it takes them out of the water, but because it allows them to perform a magic trick of disappearing behind a mirror. The fish use the fact that the surface of the sea appears as a mirror when viewed below certain angles, so predators lose track of them.

This is the same effect from below the water that makes us see the sky when looking at the surface from low angles. You can try this when swimming underwater. You'll be able to see plenty when looking anything close to straight up, but only a silvery surface when looking anything closer to horizontal. It's very common to be able to see the head and shoulders of someone standing by a swimming pool, but not their bottom half, for example.

Like butterflies, flying fish are blessed with some exquisite names: Big Raspberry, Leopardwing, Sergeant Pepper, Apache Pinkwing, Purple Haze, Violaceous Rainmaker, Pacific Necromancer . . .

Their wings often display gorgeous iridescent colors, which disappear soon after death. The reason for these colors remains a mystery, as neither courtship nor defense fully explain them; beauty that can't be explained away easily is something I am especially fond of.

They will normally only be found in seas with warm waters, more than 68 degrees Fahrenheit, and are among the most common fish in tropical surface waters.

## LOOMING

It is very common for the air just above the water to be noticeably cooler than the air a little higher up, and whenever you get bands of air of different temperatures close to each other, they will act as a lens, bending the light that we see.

When conditions are right this bending of light will make things that would normally be well beyond the horizon appear to pop into view just above the sea. The Inuit call this effect *puikkaqtuq*, which translates roughly as "popping-up," and they have used it to help them find their way and spot land from distances that would normally make it impossible. The same effect is familiar to Pacific Islanders, too, and one name for the technique there is *te kimeata*.

## BRAIDED RIVERS

We are most likely to see rivers meandering in their lower stages, but if the water runs over the right gradient and over the right type of sediment then the main channel will split into dozens of thinner ones, creating a "braided river." The thin channels in braided rivers weave their way in and out of each other and between sandbars and islands. (Braided rivers are the same effect, on a much larger scale, as the "rill marks" that we met in the Beach chapter.)

The word "braid" comes from the Middle English "brey-den," which means to move suddenly, and has its roots in the Old English "bregdan," to draw, as you would a sword. It is an apt word for channels that will switch direction almost instantly and unpredictably, abandoning an old channel and hopping to a new course regularly.

## UNDERWATER LIGHTNING

Perhaps the most mysterious water sign that I have ever heard of is *te lapa*—underwater lightning. When David Lewis was sailing with the Pacific Islanders he was told to look deep underwater, where he saw streaks and flashes of light. The Islanders were accustomed to using these light flashes to navigate, as they reliably emanated from the direction of land and could be seen best when 80 to 100 miles from an island, well out of visible range.

Any scientific explanation for these flashes is still speculative—possibly some form of bioluminescence triggered by reflected waves from the land—but it remains a mystery still to be solved.

# 19

# Uncharted Water

*An Epilogue*

W HEN SCHOLARS like David Lewis began their research on the Pacific Islands of Polynesia and Micro-
nesia in the second half of the twentieth century, a peculiar
intellectual symbiosis took place.

The Islanders had not fully appreciated the rarity of the
knowledge they had in their possession and were allowing it to
die out with the last generations who still practiced these skills.
The old ways must have seemed futile in the face of the West's
technological prowess, even before GPS. The great irony was
that this surge of Western interest rekindled an awareness
within the Islanders that they had something not just special, but unique. Nowhere else on Earth was there such a rich
repository of ancient nautical navigation techniques still alive
and in use. The Islanders had fallen into the same trap as the
West, of believing that because something has lost its value in
terms of necessity, it has lost its value altogether. A renaissance
was sparked, one culminating in organizations like the Pacific

Voyaging Society, who continue to treasure and pass on the local skills and heritage.

If I'm honest, I envied the pioneers like David Lewis, able to venture into such rich fields of untapped wisdom. This was a new nirvana for practical nautical research and I had missed the boat by a few decades, carelessly being born too late.

THERE IS A collection of stories known as the *Landnámabók*, a series of Norse accounts of Icelandic settlements in the ninth and tenth centuries. In the second chapter of an account called the *Hauksbok*, there are some intriguing references to the methods the Vikings used to find their way. The one that most piqued my curiosity was a reference to a way of maintaining latitude on a westbound voyage from Norway to Greenland.

Navigators were instructed to head far enough south that they lost sight of Iceland and saw whales, but not so far south that they lost the "coast-bound birds." This was an unmistakeable reference to an ancient use of natural navigation at sea, one that was thousands of miles from the Pacific—as well as predating the earliest of our sources there by many centuries. To my surprise, I could find no reference to anyone having investigated these methods practically. In a cynical moment I wondered if this might have been because a six-month academic posting to the golden sands of a tropical island in the Pacific may have drawn more scholars than a small boat being buffeted by icy seas. The truth is more likely that the Pacific traditions are still being practiced by living navigators, who can be interviewed, whereas the Vikings are long gone.

Academic investigations into practical Viking methods have focused disproportionately on the *solarstein* or sunstone. The sunstone is a translucent Iceland spar crystal that is claimed by some to have been used by Vikings to help them work out the direction of the sun on overcast days.

My own view is that even if the sunstone was used by Vikings, then it was symbolic—a token of a navigator's prowess and elevated status—not a serious navigation aid. There are no circumstances that I can envisage where a sunstone would prove more useful in high latitudes than the techniques in this book that the Vikings would have been familiar with, especially an intimate reading of the relationship between the water and wind, land and animals.

Contemplating these things gave rise to a now familiar feeling and reminded me of an old Norse word *aefintyr*—venture—used to convey a sense of restless curiosity.

A GOOD FRIEND, John Pahl, and I set sail from Kirkwall, in the Orkney Islands, intent on heading north. My aim was to research whether the birds, cetaceans, and other natural clues could have been used effectively by the Vikings to work out how far they were from Iceland. John was happy to help me research these methods and was as keen on the idea of a tiring holiday in the North Atlantic as most would be horrified by it.

The final preparations had made us puff. We had physically hauled almost half a ton of supplies on board the thirty-two-foot yacht. Hundreds of pounds of fuel and victuals had been passed over the guardrails. Six bright-red gas cans needed lashing at the stern and tins of soup had to be thrust into parts of the boat that are lucky if they see daylight once a year.

The lines were slipped and we headed out in a light rain, passing close to the Skerry of Vasa, where seals had found a solitary sunbeam and were having their afternoon snooze. "Skerry" is a word used in Scottish waters to refer to an isolated "drying height," a reef or rocky island, usually one that is submerged during high tides. After a brief tussle with the choppy water of the Westray Firth, off the coast of Scotland, sails were raised and we set a course just west of north. My eyes feasted on the patterns in the water, as wind, tidal currents, and landforms worked with and against each other. The sun and scudding clouds threw shadows on the shallows that led the colors of the water on a dance. As land fell astern to the south, the sea state settled. A few hours later it was evening and bright still. The sky was now filled with a humbling display of cirrus clouds.

The first night at sea, if its feeble darkness could be called that, allowed a first and, it would transpire, only, view of the stars. The Summer Triangle of Altair, Deneb, and Vega were clear and pointed south as the orange Arcturus shone brightly in the western sky. The Big Dipper and Polaris were just visible, higher in the sky than I am accustomed to seeing them. Within an hour, light levels had risen enough to hide the stars. They would not be seen again during the voyage.

On those first few days we adjusted, forgetting the luxuries of land and relearning the skills of happy offshore sailing. The art of making tea on one leg is quickly refreshed and tastes adjust to life at sea. Being out of sight of land makes a tinned curry taste like the best meal in the world.

Double-handed sailing is like single-handed sailing, but with sleep. You are alone for many hours, exchanging updates on watch handover, before heading to the bunk and welcome rest. John was spared my bubbling excitement at spotting each new pattern in the water, each shift in the waves and the resistance to change in the underlying swell. From cats' paws to the long, gentle rolls, the foam on the waves, the colors that shifted with depth and weather, the water was telling its stories and my senses fizzed.

As we approached the Faroe Islands, the sky yielded aerial clues the Vikings would have read effortlessly. We saw lumps of clouds on the horizon, directly ahead, which revealed the presence of land before we could see the islands themselves. Those "steep and high mountains" the Vikings described in the *Landnámabók* cause air to rise, which then cools, creating clouds, rain, and fog.

We watched as the clouds broke apart and allowed us to see land directly, silhouettes of shockingly stark, dark cliffs.

As we closed the Faroes there was a tidal gate we needed to make and we were running early. The tidal currents that run through the Faroe Islands have a reputation that deserves respect, especially from those in a small yacht. We hove-to when we still had about twenty miles to run. I did not want us to be forced to claw our way from much closer if the decision was to leave the Islands to our west and go around. After heaving-to, the wind backed, becoming an easterly, and picked up to a Force 5. A Force 5 inshore is near-idyllic, but it felt much more like a Force 8 in this open water with its interminable fetch, and the sea state rose to an uncomfortable level very quickly. With a second reef in the mainsail and the boat pitching irritably at each wave, the notorious tidal currents of the Faroes were foreboding.

Fortunately the sky cleared, the seas calmed a little, and within four hours we were at the mouth of Kalsoy fjord and heading into one of the eeriest sailing channels either of us have had the pleasure of encountering. Steep, dark cliffs loomed over us, then blocked much of the weak light that a low sun was able to force through the clouds.

THE WINDS ALLOWED US to push north. Nights, which became no more dramatic than a dip in the light levels, grew shorter and shorter, until they vanished altogether. Sunrise and sunset became one. We had done it, we had reached the Arctic Circle.

We turned to head west and sailed along the top of Iceland toward Greenland. Pilot whales, killer whales, and dolphins passed close by. Encountering cetaceans in a yacht is always a profound experience; there is something gripping about the way these creatures appear to simultaneously acknowledge your presence and continue their journeys in an aloof manner. One dolphin was less phlegmatic and amazed us with its acrobatics, flying high into the air.

Then the sea itself changed color, dramatically so, as we entered a spur from the East Greenland current. This flows from the far north, and in its cold, nutrient-rich waters phytoplankton thrive, turning the water a milky blue, very different from the dark-stone color of the deep waters we had been sailing through. At times we could see a stark line in the water between the two, as if at the merging of two great rivers. The current is visible from space.

The temperature plummeted and we scoured the horizon for icebergs and made a call with a satellite phone to check if any had been reported that far south during the season.

Approaching Iceland from the northwest, a spectacular landscape rose before us, with peaks from *The Lord of the Rings*. We closed on land and headed gingerly into the Jökulfirðir fjord. We sailed past the Drangajökull glacier into shallow, uncharted waters. In our world, where Google Earth can spin you close up to a mountain in the middle of Antarctica, where a satellite phone will work from pole to pole, and Wikipedia can answer questions you never thought to ask, the unknown has a strange power.

We were now sailing south, powered by the fierce Force 7 katabatic winds that rolled down off the icy mountains to our immediate west. In fog the puffins helped us picture where land was as they flew dependably toward it in their countless trains.

Reykjavik appeared, a strip of buildings dominated by the proud Hallgrímskirkja church, sandwiched between sea and icy hills. We felt torn between the desire to explore more of those wonderfully wild and beautiful seas, to travel further west or north, and the more measured decision to finish on a high note, safely. Over a thousand years ago, Eric the Red chose the former, sailing on to Greenland, but then of the twenty-four boats in his fleet, half never made it.

Unlike that Viking, we moored up and failed to repress grins as we realized that the fingers that had wandered over a chart in a London restaurant had now worked harder to forge a memorable voyage, through the Faroes, up to the Arctic Circle, and into uncharted waters. Then we did what all sensible sailors do at the end of a long voyage and talked a big game about heading out to a bar, before promptly falling asleep. Between Kirkwall and arriving at Reykjavik, over 1,000 nautical miles of ocean, we hadn't seen another sail, not a sailing vessel of any description.

THE ACADEMIC PAPER I wrote off the back of this voyage was titled "Nature's Radar." The full paper was published by the Royal Institute of Navigation's journal and is freely available from my website: naturalnavigator.com/the-library/ natures-radar-natural-navigation-research. Our voyage helped to confirm that the birds and other natural signs could be used as an effective way of gauging the distance from land in northern waters, just as the Vikings had asserted.

Shortly after the paper was published, I was done a huge favor by the military, who condensed "Nature's Radar" into a few lines that will now be found on the survival flip charts in all UK military aircraft.

> If you count more than ten birds in a random five-
> minute period you are within forty miles of land and
> if you count two or fewer then you are more than forty
> miles from land and in-between you can't be sure.

We had seen the midnight sun and sublime wildlife. I was suitably awed by whales and found myself surprised by the twitching beauty of lion's mane jellyfish. However, it was the patterns in the water and their relationship with land, sky, animals, and plants that resonated most strongly. Patterns that never stopped changing and yet were constant, ripples that ran from the puddles near my back door to the North Atlantic.

As we peered down from the small boat, surrounded by glaciers in uncharted water, and used colors to make sense of the water around us, I realized that our journey into the Arctic Circle meant I had come full circle. Every single one of the signs, clues, and patterns in this book helped on that voyage, both practically and for enjoyment. But I would have noticed or understood very few of them if I had not decided many years ago to relish studying the water much closer to home.

Given a choice between the chance to go back on such a voyage again or to know that I can see these signs all around me nearer home, it wouldn't take a minute to choose.

# Sources, Notes, and Further Reading

**STRANGE BEGINNINGS: AN INTRODUCTION**
"Water, many scientists now believe . . .": Nicholas St. Fleur, "The Water in Your Glass Might Be Older Than the Sun," *The New York Times* (Accessed 06/14/16.)
"If there is magic on this planet . . .": Loren Eiseley, todayinsci.com/E/ Eiseley_Loren/EiseleyLoren-Quotations.htm (Accessed on 06/08/16.)
Story of Abharah's Seamanship: George Hourani, *Arab Seafaring*, pp. 114–17.
*Isharat*: G. R. Tibbets, *Arab Navigation*, p. 273.
Pacific Island references: David Lewis and Stephen Thomas, *passim.*
"It really is a pretty unique set of skills . . .": voices.nationalgeographic. com/2014/03/03/hokulea-the-art-of-wayfinding-interview-with-a-master-navigator (Accessed 03/04/15.)
*Kapesani lemetau*: Stephen Thomas, *The Last Navigator*, p. 26.
*Maneaba*: David Lewis, *We, the Navigators*, p. 202.
"Do not imagine that the natives . . .": Harold Lindsay, *The Bushman's Handbook*, p. 1.
Ian Proctor: Ian Proctor, *Sailing Strategy*, p. 1.

**1. LAUNCHING**
"When there is a sudden lowering of air pressure . . .": Paul Younger, *Water*, p. 14.

## 2. HOW TO SEE THE PACIFIC IN A POND

James Cook, *A Voyage Towards the South Pole and Round the World*, Strahan and
Cadell, 1777, p. 316.

*Meaify*: Thomas, p. 78.

Marshall Islands: indebted to David Lewis, *The Voyaging Stars*, pp. 117–19.

"Like people's faces": Lewis, *WTN*, p. 132.

Hipour, Big Wave, and Big Bird: ibid., p. 130.

## 3. LAND RIPPLES

"In 1885 the South Australian Government": Lindsay, p. 20.

The black poplar is the most endangered native timber tree:
woodlandtrust.org.uk/visiting-woods/trees-woods-and-wildlife/british-trees/
native-trees/black-poplar (Accessed 05/11/15.)

Rim lichen: ocean.si.edu/blog/seaside-lichens (Accessed 06/14/16.)

Bourne, burn, brook, strath, gill, aber: Nigel Holmes and Paul Raven, *Rivers*,
pp. 18–19.

## 4. THE NOT-SO-HUMBLE PUDDLE

"Let the reader when he next receives a cup of tea . . .": A. M. Worthington, *A
Study of Splashes*, p. 30.

*Vryses*: Adam Nicolson, p. 61.

"A window to another dimension": blog.eyeem.com/2012/07/how-to-shoot-
puddleography (Accessed on 04/15/15.)

## 5. RIVERS AND STREAMS

There was an attempt in the 1920s to classify river stages according to the fish:
Holmes and Raven, p. 123.

A stream is just a river you can step over: Holmes and Raven, p. 15.

Rain before St. Valentine's Day: Simon Cooper, *Life of a Chalkstream*, p. 118.

Style of bridges: Holmes and Raven, p. 65.

"A nomad I will remain for life . . .": Gisela Brinker-Gabler (Ed.), *Encountering
the Other(s): Studies in Literature, History, and Culture*, State University of New
York Press, 1995, p. 297.

Habit in places that flood that stretches back to Ancient Egypt: Daniel
Kahneman, *Thinking Fast and Slow*, Farrar, Straus and Giroux, 2011, p. 137.

Reed canary grass: Holmes and Raven, p. 194.

If water table reaches the ground: Philip Ball, $H_2O$, p. 40.

It is a clue to the height of the summer water table around you: Younger, p.
24.

Snails clinging to the surface film: Cooper, p. 215.

Pioneer species like hemp agrimony (found in the eastern US), willowherbs, and young willow trees: Bellamy, p. 140.

"the pitch of which is high above . . .": Chris Watson, in Jeff Barret, Robin Turner, and Andrew Walsh (Eds.), *Caught by the River*, p. 63.

If the body is thicker than a matchstick, it's a dragonfly: David Bellamy, *The Countryside Detective*, p. 136.

Eel migrations are sensitive to water temperature, moon phase, and even atmospheric pressure: int-res.com/articles/meps2002/234/m234p281.pdf (Accessed 07/21/15.)

Birds like cormorants prefer to fish: John Pahl, private conversation.

Black Tailing: Holmes and Raven, p. 124.

"Cowbelly": my thanks to the outdoor photographer and author Dominick Tyler for making me aware of this phenomenon during a joint talk we gave in the Cotswolds.

The island is behaving like a thumb over the tap: Proctor, p. 10.

"I am an old man now, and when I die and go to Heaven . . .": from Jonathan Raban, *Passage to Juneau*, p. 291.

Granite Eddy: Rebecca Lawton, *Reading Water*, p. 46.

"On the Colorado, eddies reign supreme . . .": ibid., p. 45.

Leonardo da Vinci was entranced: Philip Ball, *Flow*, p. 10.

"Big whirls have little whirls . . .": ibid., p. 175.

A riffle-pool sequence for every stretch of river that is five times its width: Holmes and Raven, p. 91.

## 6. THE RISE

Johnson vs. Davy: Holmes and Raven, pp. 273–4.

Dry fly-fishing may date . . . Victorians: Cooper, pp. 7 and 31.

"The key to it all is thought . . .": Brian Clarke, *The Pursuit of Stillwater Trout*, pp. 12 and 16.

Yellow star thistle and zebra mussels: fws.gov/invasives (Accessed 06/14/16.)

Without a stomach: Cooper, p. 167.

Polarized sunglasses . . . wide-brimmed hat . . . shady to bright: John Goddard and Brian Clarke, *Understanding Trout Behavior*, pp. 19–24.

"That moment when a dimple radiates across . . .": Cooper, pp. 160–1.

G. E. M. Skues rise types: Kenneth Robson, p. 35, taken from *The Way of a Trout with a Fly*.

S-bend in reflections: Goddard and Clarke, p. 57.

*chink of light*: Cooper, p. 112.

Victorians classified several different species of brown trout: Holmes and Raven, p. 259.

Clarke rise types: Clarke, pp. 110–21.

## 7. THE LAKE

6,750 litres: Heather Angel and Pat Wolseley, *The Family Water Naturalist*, p. 10. Our sense of smell takes a different route . . .":

Wallace Nichols, *Blue Mind*, p. 95.

Ignore the faintest whiff of a rat: Tom Cunliffe, *Inshore Navigation*, p. 64.

From 50 tards deep in very clear lakes to only 20 inches in very silty waters . . .: Mary Burgis and Pat Morris, *The Natural History of Lakes*, p. 25.

You can actually create your own thermocline experiment in the kitchen if you want to: Angel and Wolseley, p. 11.

A sound barrier that blocks sonar—it is used by military submarines to hide: Terry Breverton, *Breverton's Nautical Curiosities*, p. 351.

Whale communications: natgeotv.com.au/tv/kingdom-of-the-blue-whale/ blue-whales-and-communication.aspx (Accessed 03/26/15.)

Wind around obstacles: David Houghton and Fiona Campbell, *Wind Strategy*, pp. 62–3. Eddies: Proctor, pp. 106–7.

David Lewis and Iotiebata: Lewis, *The Voyaging Stars*, p. 115.

## 8. THE COLOR OF WATER

Celts: glasto: en.wikipedia.org/wiki/Green#Languages_where_green_and_ blue_are_one_color (Accessed 10/22/15.) They have even worked out its wavelength: 480 nanometers: David Lynch and William Livingston, *Color and Light in Nature*, p. 66.

"There's nothin' in the world can be easier, when you've learned your lesson, than to pick your way about . . .": Brian Fagan, *Beyond the Blue Horizon*, p. 200, quoting from *The Baltic and the North Seas* by Merja-Liisa Hinkkanen and David Kirby.

The Amazon is a yellow color in places and near Manaus in Brazil: Lynch and Livingston, p. 67.

Amazingly, scientists believe that our seas may be the only place in the universe where water blows over open water to create foamy waves: Sidney Perkowitz, *Universal Foam*, p. 131.

One of the biggest examples of this is the Mediterranean: epistimograph .blogspot.co.uk/2011/04/blue-waters-of-mediterranean.html (Accessed 05/15/15.)

Both the Greek and Roman goddesses of love, Aphrodite and Venus, sprang from the foam of the sea . . . Botticelli . . . : Perkowitz, pp. 4–5.

Foam is made of tiny pockets of air surrounded by water and clouds are made of . . . But look closely enough at foam: Lynch and Livingston, p. 92.

Indigo blue to greenish blue (1–5 FU scale): citclops.eu/water-colour/ measuring-water-colour (Accessed 05/16/15.)

## 9. LIGHT AND WATER

Moon circles: Lynch and Livingston, p. 82.

The prime example of this is a bridge with pillars: ibid., p. 99.

Mathematicians have worked out that this is a fairly exact science: Adam, p. 138.

Rectangular grids of bright, white dots: epod.usra.edu/blog/2014/08/capillary-waves.html (Accessed on 05/27/15.)

Scrooge: telegraph.co.uk/news/weather/11286360/God-or-Scrooge-Mysterious-face-spotted-in-the-waves.html (Accessed on 05/27/15.)

Your shadow may have an orange-hued: John Naylor, *Out of the Blue*, p. 46.

Aureole effect: Lynch and Livingston, p. 260.

## 10. THE SOUND OF WATER

Sources are mainly from the village of Eyam itself, but also: peaklandheritage.org.uk/index.asp?peakkey=40402121 (site closed but archive available).

It turns out that the long ribbon-like leaves of corn and wheat scatter sound surprisingly effectively: Keith Attenborough, Kai Ming Li, and Kirill Horoshenkov, *Predicting Outdoor Sound*, CRC Press, 2006, p. 320.

Schiphol Airport story: wired.com/2014/06/airport-schiphol (Accessed 03/24/15.)

Battle of Iuka: Peter Cozzens, *The Darkest Days of War*, University of North Carolina Press, 2006.

"I have a friend who lives on the south bank of a broad part of the Thames in London . . .": John Pahl, personal conversation.

Chukchi hunters: rtd.rt.com/films/i-am-hunter (Accessed 01/10/15.)

## 11. READING WAVES

The motion at the top of the wave is slightly faster than the motion at the bottom: Drew Kampion, *The Book of Waves*, p. 38.

As the ocean scientist Willard Bascom put it: Willard Bascom, *Waves and Beaches*, p. 11.

It is popularly supposed that every seventh wave will be bigger . . . approximately 1 in 2000 according to the ocean scientists: Houghton and Campbell, p. 69.

The speed in knots of a wave at sea will be its period in seconds multiplied by 3: Proctor, p. 50.

On September 8, 1900, locals remarked on the extraordinary swell hitting the beach in Galveston: Carl Hobbs, *The Beach Book*, p. 38.

A submarine need only descend to 150 yards: Pretor-Pinney, p. 31.

"There is no such thing as an atheist transatlantic sailor": Sarah Morison, personal conversation, quoting her brother.

"There on the coast a haven lies, named for Phorcys . . .": Homer, *Odyssey*, book 13, lines 109–14, p. 289, from Fagan, p. 89.

Long Beach, California, O'Brien, wave refraction lens: Bascom, p. 74.

Tree and DVD diffraction examples: en.wikipedia.org/wiki/Diffraction (Accessed on 06/09/15.)

Wave diffraction past sea wall and diagram: Proctor, p. 72.

The Fawa'id, but with a more unwieldy full title, *Kitab al-Fawa'id fi usul 'ilm al-bahr wa'l-qawa'id*, which translates as: Tibbets, pp. 25 and 252.

A debate rumbles on about whether there are three or four types of breaking wave, which is a bit of nonsense really: Gavin Pretor-Pinney, *The Wavewatcher's Companion*, p. 38.

An onshore wind will cause waves to break earlier, in deeper water . . . : Scott Douglass, "Influence of Wind on Breaking Waves": cedb.asce.org/cgi/ WWWdisplay.cgi?68193 (Accessed 06/18/15.)

If you ever want to gauge the height of breaking waves from a beach: Bascom, p. 172.

1834, the same year that the first practical electric motor was invented: en.wikipedia.org/wiki/Timeline_of_ historic_inventions (Accessed 10/26/15.)

## 12. THE OMANI DELIGHT: AN INTERLUDE
Personal recollections.

## 13. THE COAST
Only 0.04 percent . . .: water.usgs.gov/edu/gallery/global-water-volume.html (Accessed 01/08/16).

The winds that reached Tonga from the northwest were warm and moist . . . : Lewis, *The Voyaging Stars*, p. 76.

Homer's Nestor and Eurymedon: Jamie Morton, *The Role of the Physical Environment in Ancient Greek Seafaring*, p. 52.

Nainoa Thompson: Will Kyselka, *An Ocean in Mind*, p. 26.

Ammassalik wooden maps: John Pahl, personal correspondence.

Elizabeth I . . . was made a criminal offense: Bella Bathurst, *The Lighthouse Stevensons*, p. 7.

Parafungen: Lewis, *The Voyaging Stars*, p. 138.

*Pookof*: Thomas, p. 258 and *passim*.

A charming thirteenth-century account . . . : Ibn Mujawir, *A Traveller in Thirteenth-Century Arabia: Ibn al-Mujawir's Tarikh al-Mustabsir*, trans. by G. Rex Smith (London: Hakluyt Society, 2008), p. 264. My thanks to Eric Staples for drawing my attention to this wonderful example.

Robert Stevenson . . . fish, weather: Bathurst, p. 94.

Ironically, since rip currents can have a smoothing effect on the waves, they can attract swimmers: Bascom, p. 170.

## 14. THE BEACH

They can flow at eight feet per second, faster than any swimmer: Lisa Woollett, *Sea and Shore Cornwall*, p. 87.

"Mariners are advised that a bathymetric survey of Chichester Bar . . .": Email from Chichester Harbour Conservancy, 12/24/14.

"Interference ripples," "flaser ripples" . . . : coastalcare.org/educate/ exploring-the-sand (Accessed on 06/11/15.)

More like thin diamonds . . . will begin to seep back out when the tide retreats: Bascom, p. 206.

If you rest a finger on the domes they will collapse: ibid., p. 210.

Spannel: Woollett, p. 19.

You never see just one groin . . . : Hobbs, p. 151.

Scilly Isles . . . are famous for their white sands: Clifford and King, p. 14.

Studies have revealed that each patch of sand in the world is unique: Clifford and King, p. 35.

Fishermen along Chesil Beach: ibid., p. 13.

"Sing of the shore": Woollett, p. 19.

"People are always trying to find the next strongest thing . . .": bbc.co.uk/ news/science-environment-31500883 (Accessed on 06/12/15.)

Gray cases are a sign . . . : bbc.co.uk/insideout/south/series6/beachcombing .shtml (Accessed on 06/15/15.)

Sailors did once use them for washing: Woollett, p. 17.

Wrecks form extraordinarily rich breeding: ibid., p. 15.

Periwinkles: ibid., p. 73.

*Silvetia*, bladder, and saw (or serrated) wracks are considerate: Tristan Gooley, *The Lost Art of Reading Nature's Signs*, p. 286.

Driftwood: Woollett, p. 23.

Gilbert Islands weather lore: Lewis, *The Voyaging Stars*, pp. 124–5.

## 15. CURRENTS AND TIDES

"What has rather complicated today's" . . . Marcus Vergette, the Bell's sculptor, told the bemused audience . . . : John Pahl, personal correspondence.

In May 1990 a storm . . . eleven years later: Woollett, p. 62.

Stir tea by blowing: Ball, *Flow*, p. 39.

Wind-driven currents to struggle to get much above 2 percent of the wind speed: Proctor, p. 16.

A 10-knot wind blowing over warm water that is only one yard deep: Houghton and Campbell, p. 65.

The global average is only about half a knot: David Burch, *Emergency Navigation*, p. 130.

A current is moving water at 2 knots on a still air day: Proctor, p. 47.

Currents will have a small effect on the wavelength and height of any waves: Houghton and Campbell, p. 68.

Nainoa Thompson, of the Polynesian Voyaging Society . . . : Kyselka, p. 149.

Expert racing sailors will scour the water: Proctor, p. 74.

That Wellington boots that were lost by fishermen in the North Sea: Woollett, p. 95.

The line of beach known as *Ka-Milo-Pae-Ali'i*, which translates roughly as "the twisting water washes ashore royalty" . . . : Curtis Ebbesmayer and Eric Scigliano, *Flotsametrics and the Floating World*, p. 198.

In 1973 he became one of the first people to successfully swim the eighteen miles around the Channel Island of Sark: Roger Deakin, *Waterlog*, p. 36.

Fastest current, Bodo: Pretor-Pinney, p. 231.

From the Warao people of the Orinoco Delta: John Pahl, personal correspondence.

In the Arctic, the Iglulingmiut, an Inuit group, detect which way the currents are flowing by looking at the *qiqquaq* fronds: John Macdonald, *The Arctic Sky*, p. 183.

From 1833, when the first tide tables were produced by the Admiralty: James Greig McCully, *Beyond the Moon*, p. 6.

Alexander the Great was mystified by the tides and even Galileo misunderstood them: ibid., pp. 1–4.

"The ocean just sat there": Lewis, *The Voyaging Stars*, p. 116 and en.wikipedia .org/wiki/Battle_of_Tarawa (Accessed 06/29/15.)

"Initially, the idea on Wednesday had been to deliberately refloat the ship . . .": theguardian.com/uk-news/2015/jan/08/car-carrier-beached-solent-sandbank-refloats-itself (Accessed 06/29/15.)

On February 5, 2004, twenty-one Chinese laborers drowned while picking cockles . . . : en.wikipedia.org/wiki/2004_Morecambe_Bay_cockling_ disaster (Accessed 06/29/15.)

"Reclaim the Beach" parties along the Thames: Clifford and King, p. 14.

If you are lucky you may witness shipwrecks or petrified forests: Woollett, p. 74.

There is no mention of tides in the Bible: Pretor-Pinney, p. 251.

"Now the whole country of India has very many rivers . . .": legacy.fordham .edu/halsall/ancient/periplus.asp (Accessed 06/24/15.)

If you see the height and pattern of tides doing different things on different days: Greig McCully, p. 57.

Among the First Nation people of British Columbia: Raban, p. 224.

If the tide is much higher than expected or predicted . . . then this may be a warning of strong winds out to sea: Proctor, p. 24.

The farther from the sea you go, the later the tide will turn on average . . . the bizarre consequence of a river temporarily running in different directions: Proctor, p. 22.

The ebb is usually stronger than the flood . . . : Cunliffe, p. 41.

A friend reports: John Pahl, personal conversation. "They are aware of these dangers beforehand, and whenever they come close to the headland . . .": Aelianus, translated by A. F. Schofield, quoted in Morton, p. 41.

The National Oceanic and Atmospheric Administration . . . The oceanographer, Dr. Arthur Doodson, believes there are a total of 396 factors at play: Greig McCully, p. 11.

## 16. WATER AT NIGHT

By the 1830s, more than two ships a day were getting wrecked around the British coastline: Bathurst, *The Lighthouse Stevensons*, p. 10.

A swarm of them swamped a salmon farm in Northern Ireland: en.wikipedia .org/wiki/Pelagia_noctiluca (Accessed on 06/02/15.)

## 17. SHIPWATCHING

Ibn-Jubayr: "So we were able to see the art of these captains and sailors . . .": Hourani, p. 122

"'Tis the aftermost grinder aloft . . .": Peter Kemp, *Oxford Companion to Ships and the Sea*, preface.

Some have observed wryly that we can gauge the global economy by studying these lines and that shortly after 2008 they leapt out of the water.: Horatio Clare, *Down to the Sea in Ships*, p. 26.

## 18. RARE AND EXTRAORDINARY

Kelvin waves: Greig McCully, p. 85.

In 2004, for the first time accurate measurements were made of a tsunami's height in this early stage: www.noaanews.noaa.gov/stories2005/s2365.htm (Accessed 07/03/15.)

230,000 people in fourteen countries: en.wikipedia.org/wiki/2004_Indian_ Ocean_earthquake_and_tsunami (Accessed 07/03/15.)

The Moken sea gypsies: cbsnews.com/news/sea-gypsies-saw-signs-in-the-waves (Accessed 07/06/15.)

When there is no other explanation for a wave . . . : Grieg McCully, p. 101.

Tidal bore: Kampion, p. 39.

Cook Inlet: nationalgeographic.org/encyclopedia/tidal-bore (Accessed 06/14/16.)

The *Glamorgan*: Kampion, p. 43.

Old Sow: oceanservice.noaa.gov/facts/old-sow.html (Accessed 06/14/16.)

Slippery water: Houghton and Campbell, p. 66.

Flying Fish: this section is indebted to *The Amazing World of Flyingfish* by Steve Howell.

*Puikkaqtuq*, which translates roughly as "popping-up": Macdonald, p. 185.

*Te kimeata*: Lewis, *The Voyaging Stars*, p. 123. Braided Rivers: Lawton, pp. 19–21.

Underwater Lightning: Lewis, *The Voyaging Stars*, pp. 48–9.

## UNCHARTED WATER: AN EPILOGUE

*Aefintyr*: Fagan, p. xvii.

Tristan Gooley, "Nature's Radar," *Journal of Navigation*, 66, 2013, pp. 161–79.

"the survival flip charts in all UK military aircraft": JSP 374 Aircrew SERE flip card, with thanks to John Hudson.

# Selected Bibliography

Angel, Heather, and Pat Wolseley, *The Family Water Naturalist*, Michael Joseph, 1982.

Ball, Philip, *H₂O*, Phoenix, 1999.

Ball, Philip, *Branches*, Oxford University Press, 2009.

Ball, Philip, *Flow*, Oxford University Press, 2009.

Barkham, Patrick, *Coastlines*, Granta, 2015.

Barrett, Jeff, Robin Turner, and Andrew Walsh, *Caught by the River*, Cassell Illustrated, 2009.

Bartholomew, Alick, *The Story of Water*, Floris Books, 2010.

Bascom, Willard, *Waves and Beaches*, Anchor Books, 1964.

Bathurst, Bella, *The Lighthouse Stevensons*, Harper Perennial, 2005.

Bathurst, Bella, *The Wreckers*, Harper Perennial, 2006.

Bellamy, David, *The Countryside Detective*, Reader's Digest Association, 2000.

Breverton, Terry, *Breverton's Nautical Curiosities*, Quercus, 2010.

Bruce, Peter, *Heavy Weather Sailing*, Adlard Coles Nautical, 1999.

Burch, David, *Emergency Navigation*, McGraw-Hill, 2008.

Burgis, Mary, and Pat Morris, *The Natural History of Lakes*, Cambridge University Press, 1987.

Clare, Horatio, *Down to the Sea in Ships*, Vintage, 2015.

Clarke, Brian, *The Pursuit of Stillwater Trout*, A & C Black, 1975.

Clifford, Sue, and Angela King, *Journeys Through England in Particular: Coasting*, Saltyard, 2013.

Cooper, Simon, *Life of a Chalkstream*, William Collins, 2014.

Cox, Lynne, *Open Water Swimming Manual*, Vintage, 2013.

Cunliffe, Tom, *Inshore Navigation*, Fernhurst Books, 1987.

Deakin, Roger, *Waterlog*, Vintage, 2000.

Ebbesmeyer, Curtis, and Eric Scigliano, *Flotsametrics and the Floating World*, HarperCollins, 2010.

Evans, I. O., *Sea and Seashore*, Frederick Warne & Co., 1964.

Fagan, Brian, *Beyond the Blue Horizon*, Bloomsbury, 2012.

Ferrero, Franco, *Sea Kayak Navigation*, Pesda Press, 2009.

Gatty, Harold, *The Raft Book*, George Grady, 1944.

Goddard, John, and Brian Clarke, *Understanding Trout Behavior*, Lyons Press, 2001.

Goodwin, Ray, *Canoeing*, Pesda Press, 2011.

Gooley, Tristan, *The Natural Navigator*, The Experiment, 2011.

Gooley, Tristan, *The Natural Explorer*, Sceptre, 2012.

Gooley, Tristan, *How to Connect with Nature*, Macmillan, 2014.

Gooley, Tristan, *The Lost Art of Reading Nature's Signs*, The Experiment, 2015.

Gooley, Tristan, "Nature's Radar," *Journal of Navigation*, 66, 2013, pp. 161–79, doi:10.1017/S0373463312000495.

Greig McCully, James, *Beyond the Moon*, World Scientific Publishing, 2006.

Hill, Peter, *Stargazing*, Canongate Books, 2004.

Hobbs, Carl, *The Beach Book*, Columbia University Press, 2012.

Holmes, Nigel, and Paul Raven, *Rivers*, British Wildlife Publishing, 2014.

Houghton, David, and Fiona Campbell, *Wind Strategy*, Fernhurst Books, 2012.

Hourani, George, *Arab Seafaring*, Princeton University Press, 1995.

Howell, Steve, *The Amazing World of Flyingfish*, Princeton University Press, 2014.

Humble, Kate, and Martin McGill, *Watching Waterbirds*, A & C Black, 2011.

Huth, John, *The Lost Art of Finding Our Way*, Belknap Press, 2013.

Kampion, Drew, *The Book of Waves*, Roberts Rinehart, 1991.

Karlsen, Leif, *Secrets of the Viking Navigators*, One Earth Press, 2003.

Kemp, Peter, *The Oxford Companion to Ships and the Sea*, Oxford University Press, 1979.

Kyselka, Will, *An Ocean in Mind*, University of Hawaii Press, 1987.

Lawton, Rebecca, *Reading Water*, Capital Books, 2002.

Lewis, David, *The Voyaging Stars*, Fontana, 1978.

Lewis, David, *We, the Navigators*, University of Hawaii Press, 1994.

Lindsay, Harold, *The Bushman's Handbook*, Angus Robertson, 1948.

Lynch, David, and William Livingston, *Color and Light in Nature*, Cambridge University Press, 1995.

MacDonald, John, *The Arctic Sky*, Royal Ontario Museum, 1998.

Morton, Jamie, *The Role of the Physical Environment in Ancient Greek Seafaring*, Brill, 2001.

Naylor, John, *Out of the Blue*, Cambridge University Press, 2002.

Nichols, Wallace, *Blue Mind*, Little, Brown, 2014.

Nicolson, Adam, *Why Homer Matters*, Picador, 2014.

Pearson, Malcolm, *Reed's Skipper's Handbook*, Reed Thomas Publications, 2000.

Perkowitz, Sidney, *Universal Foam*, Vintage, 2001.

Plass, Maya, *RSPB Handbook of the Shore*, A & C Black, 2013.

Pretor-Pinney, Gavin, *The Wavewatcher's Companion*, Bloomsbury, 2010.

Proctor, Ian, *Sailing Strategy*, Adlard Coles Nautical, 2010.

Raban, Jonathan, *Passage to Juneau*, Picador, 1999.

Rex Smith, G., *A Traveller in Thirteenth-Century Arabia: Ibn al-Mujawir's Tarikh al Musrabsir*, Hakluyt Society, 2008.

Robson, Kenneth, *The Essential G. E. M. Skues*, A & C Black, 1998.

Severin, Tim, *The Ulysses Voyage*, Book Club Associates, 1987.

Sharp, Andrew, *Ancient Voyagers of the Pacific*, Penguin, 1957.

Steers, J. A., *The Sea Coast*, Collins, 1962.

Sterry, Paul, *Pond Watching*, Hamlyn, 1983.

Taylor, E. G. R., *The Haven-Finding Art*, Hollis & Carter, 1956.

Thomas, David, and David Bowers, *Introducing Oceanography*, Dunedin Academic Press, 2012.

Thomas, Stephen, *The Last Navigator*, Random House, 1987.

Tibbets, G. R., *Arab Navigation*, The Royal Asiatic Society of Great Britain, 1971.

Tyler, Dominick, *Uncommon Ground*, Guardian Books, 2015.

Walker, Stuart, *Wind and Strategy*, W. W. Norton & Co., 1973.

Woollett, Lisa, *Sea and Shore Cornwall*, Zart Books, 2013.

Worthington, A. M., *A Study of Splashes*, Longmans, Green & Co., 1908.

Yates, Chris, *How to Fish*, Penguin, 2006.

Younger, Paul, *Water*, Hodder & Stoughton, 2012.

# Acknowledgments

THERE IS A LOT MORE to water than initially meets the eye and there is a lot more to authorship than the name on the cover. I once read, I forget where, that there are thirty-eight stages between the submission of a "completed" manuscript and the publication of a book. I laugh at the impossibility of that figure while writing a book and then nod at its wisdom come publishing day.

The team behind this book did a brilliant job of chaperoning me through those stages, and I would like to thank Maddy Price, Neil Gower, Rebecca Mundy, Caitriona Horne, and the team at Sceptre for their invaluable help and hard work. But, any errors in this book are my own, as indeed are any follies. There are too many people I would like to thank for helping me explore the subject of this book to do everyone justice. However, I would like to thank the following for going out of their way to help over the past couple of years: John Pahl, Eric Staples, Stuart Crofts, and my sister, Siobhan Machin, thank you.

I would also like to thank all those who support my work in a low profile but invaluable way: those that come on courses, those who buy my books, those that write to me with gems from near and far, those that spread the word about the work I do and therefore make it possible. You know who you are and I salute you!

For making the US edition of this book possible, I would like to thank Nicholas Cizek, Jeanne Tao, Sarah Smith, Jennifer Hergenroeder, and the rest of the team at The Experiment. I would also like to thank my publisher, Rupert Lancaster, for commissioning this book in the fullest sense. It was his enthusiasm and belief that this book needed to be written and by me that overcame my slight fear of the work that would be needed. It is unlikely that it would exist without that encouragement. I would like to thank both Rupert and my agent, Sophie Hicks, for their limitless help, patience, and support from commission to publication.

My final thanks goes to my family for tolerating such a curious creature in their midst. Not long ago, I stopped by a river to point something out to my younger son. He shook his head, sighed, and said, "Oh no . . . Not this guy again!"

# Picture Acknowledgments

Illustrations ©Neil Gower
Photographs ©Tristan Gooley
From: *The Splash of a Drop* by Prof. A. M. Worthington, London 1895: p. 57
Landscape by Moonlight c. 1635–40 by Peter Paul Rubens © Samuel
    Courtauld Trust, The Courtauld Gallery London/Bridgeman Images:
    p. 145
©Top Photo Corporation/Shutterstock: p. 153
©ChrisVanLennepPhoto/Shutterstock: p. 202
Graph redrawn from Haver, Sverre (2003). *Freak Wave Event at Draupner Jacket
    January 1, 1995*: p. 342

# Index

Page references in *italics* indicate photos/illustrations.

# About the Author

TRISTAN GOOLEY, writer, navigator, and explorer, is also the author of *The Lost Art of Reading Nature's Signs* and *The Natural Navigator*. Through his journeys, teaching, and writing, he has pioneered a renaissance in the rare art of natural navigation. Tristan has led expeditions in five continents and climbed mountains in Europe, Africa, and Asia. He has explored close to home and walked with and studied the methods of tribal peoples in some of the remotest regions on Earth, and is the only living person to have both flown solo and sailed single-handedly across the Atlantic.